CAMBRIDGE CHECKPOINTS 2018–2022

Year 11 (Stage 6) Biology

- Sample examination questions
- Questions arranged by topic
- Suggested responses to questions

Harry Leather & Jan Leather

CAMBRIDGE
UNIVERSITY PRESS

Shaftesbury Road, Cambridge CB2 8EA, United Kingdom

One Liberty Plaza, 20th Floor, New York, NY 10006, USA

477 Williamstown Road, Port Melbourne, VIC 3207, Australia

314–321, 3rd Floor, Plot 3, Splendor Forum, Jasola District Centre, New Delhi – 110025, India

103 Penang Road, #05–06/07, Visioncrest Commercial, Singapore 238467

Cambridge University Press is part of Cambridge University Press & Assessment, a department of the University of Cambridge.

We share the University's mission to contribute to society through the pursuit of education, learning and research at the highest international levels of excellence.

www.cambridge.org

First published 2018
20 19 18 17 16 15 14 13 12 11 10 9 8 7 6

Printed in Australia by Ligare Book Printers.

A catalogue record for this book is available from the National Library of Australia at www.nla.gov.au

ISBN 978-1-108-43532-1 Paperback

Additional resources for this publication at www.cambridge.edu.au / GO

Cambridge University Press & Assessment acknowledges the Australian Aboriginal and Torres Strait Islander peoples of this nation. We acknowledge the traditional custodians of the lands on which our company is located and where we conduct our business. We pay our respects to ancestors and Elders, past and present. Cambridge University Press & Assessment is committed to honouring Australian Aboriginal and Torres Strait Islander peoples' unique cultural and spiritual relationships to the land, waters and seas and their rich contribution to society.

Contents

Chapter 7
Sample Examination Paper

Chapter 8
Suggested Answers: Sample Examination Paper

Introduction
Preliminary Biology End-of-Year Examination

This book contains relevant questions useful for revising Preliminary Biology in preparation for an end-of-year examination. It includes questions that cover the material in each of the four modules: 'Cells as the Basis of Life', 'Organisation of Living Things', 'Biological Diversity' and 'Ecosystem Dynamics'. The questions are grouped into chapters that reflect the course as described in the Syllabus. The headings are similar to those that are used in the Preliminary Biology Syllabus. The questions for the 'Biology Skills' module are spread through chapters 2 to 5 which look at the other four core areas.

Each of the chapters consists of three sections. The first section is a list of key words. You should be familiar with all these words, and be able to define and use them in context. The second section contains content review questions. These look at major ideas that you should understand and aim to develop your knowledge base in Biology.

The final section contains the most important questions - sample examination questions. The questions in the sample end-of-year examination will be in two parts. The sample paper is structured in the same way as the HSC external examination.

In **Section 1**, 20 marks are allotted to multiple-choice questions where you need to select the correct response. The remaining 80 marks in **Section 2** of the paper contain short-answer questions. In the Preliminary course you are expected to cover all modules – all modules are compulsory.)

The short-answer questions start with a paragraph of information. This is used as a focus for examining particular ideas. The examination does more than just test your knowledge. It aims to test your ability to analyse a new situation and apply the concepts you have learned. This is something with which students initially have difficulty. The good news is that with practice, you will begin to tackle the questions asked with more confidence.

Chapters 2 to 5 contain questions that are of the same style as questions in the external HSC Biology examination papers. The marks allotted are shown next to each of these questions. Answer these questions during the semester as you complete a topic.

Chapter 6 contains answers to the review questions and the examination style questions. Chapter 7 contains a whole sample examination paper. We would suggest that you attempt this paper in your preparation leading up to the end of year examination. Chapter 8 contains suggested answers to the sample examination paper.

Introduction
Preliminary Biology End-of-Year Examination

Chapter 1
What do I have to know?

The end of year examination will be 3-hour, internally administered, written examination (length may vary depending on a school's assessment policies) of the four compulsory modules. Your teacher will have pointed out the current BOS Preliminary Biology Syllabus and the relevant contents. If she/he has not, you should ask how you can obtain a copy and become familiar with its contents. Using current biological articles and issues, the examination assesses a student's ability to:

- recall facts, definitions and examples;
- explain biological concepts, principles and processes;
- apply understanding of concepts and processes to unfamiliar situations;
- analyse the relationship of data to concepts and principles;
- evaluate experimental procedures;
- solve problems or make predictions or judgements based on the synthesis of a number of ideas.

The course content is organised into four distinct modules and teachers will use different contexts to teach the course. Apart from the four modules students are expected to have covered the **'Working Scientifically Skills'** content shown below and as described in the NESA Biology Syllabus. These skills are to be covered over the two years of the Biology Syllabus.

Questioning and Predicting

- develop and evaluate inquiry questions and hypotheses to identify a concept that can be investigated scientifically, involving primary and secondary data
- modify questions and hypotheses to reflect new evidence

Planning Investigations

- assess risks, consider ethical issues and select appropriate materials and technologies when designing and planning an investigation
- justify and evaluate the use of variables and experimental controls to ensure that a valid procedure is developed that allows for the reliable collection of data
- evaluate and modify an investigation in response to new evidence

Conducting Investigations

- employ and evaluate safe work practices and manage risks use appropriate technologies to ensure and evaluate accuracy
- select and extract information from a wide range of reliable secondary sources and acknowledge them using an accepted referencing style

Processing Data and Information

- select qualitative and quantitative data and information and represent them using a range of formats, digital technologies and appropriate media
- apply quantitative processes where appropriate
- evaluate and improve the quality of data

Analysing Data and Information

- derive trends, patterns and relationships in data and information
- assess error, uncertainty and limitations in data
- assess the relevance, accuracy, validity and reliability of primary and secondary data and suggest improvements to investigations

Problem Solving

- use modelling (including mathematical examples) to explain phenomena, make predictions and solve problems using evidence from primary and secondary sources
- use scientific evidence and critical thinking skills to solve problems

Communicating

- select and use suitable forms of digital, visual, written and/or oral forms of communication
- select and apply appropriate scientific notations, nomenclature and scientific language to communicate in a variety of contexts
- construct evidence-based arguments and engage in peer feedback to evaluate an argument or conclusion

Preliminary Biology Modules

The content focus and content areas as specified in the Preliminary Biology Syllabus are:

Module 1 Cells as the basis of life

The NESA Syllabus gives the following content focus for this module:

'Cells are the basis of life. They coordinate activities to form colonial and multicellular organisms. Students examine the structure and function of organisms at both the cellular and tissue levels in order to describe how they facilitate the efficient provision and removal of materials to and from all cells in organisms. They are introduced to and investigate biochemical processes through the application of the Working Scientifically skills processes.

Students are introduced to the study of microbiology and the tools that scientists use in this field.

These tools will be used throughout the course to assist in making predictions and solving problems of a multidisciplinary nature.

The content summary of what students are expected to cover in this module are as follows:

Cell Structure

- investigate different cellular structures, including but not limited to:
  - examining a variety of prokaryotic and eukaryotic cells
  - describe a range of technologies that are used to determine a cell's structure and function
- investigate a variety of prokaryotic and eukaryotic cell structures, including but not limited to:
  - drawing scaled diagrams of a variety of cells
  - comparing and contrasting different cell organelles and arrangements
  - modelling the structure and function of the fluid mosaic model of the cell membrane

Cell Function

- investigate the way in which materials can move into and out of cells, including but not limited to:
  - conducting a practical investigation modelling diffusion and osmosis
  - examining the roles of active transport, endocytosis and exocytosis
  - relating the exchange of materials across membranes to the surface-area-to-volume ratio, concentration gradients and characteristics of the materials being exchanged
- investigate cell requirements, including but not limited to:
  - suitable forms of energy, including light energy and chemical energy in complex molecules
  - matter, including gases, simple nutrients and ions
  - removal of wastes
- investigate the biochemical processes of photosynthesis, cell respiration and the removal of cellular products and wastes in eukaryotic cells
- conduct a practical investigation to model the action of enzymes in cells
- investigate the effects of the environment on enzyme activity through the collection of primary or secondary data

Module 2 Organisation of Living Things

The NESA Syllabus gives the following content focus for this module:

'Multicellular organisms typically consist of a number of interdependent transport systems that range in complexity and allow the organism to exchange nutrients, gases and wastes between the internal and external environments. Students examine the relationship between these transport systems and compare nutrient and gas requirements.

Models of transport systems and structures have been developed over time, based on evidence gathered from a variety of disciplines. The interrelatedness of these transport systems is critical in maintaining health and in solving problems related to sustainability in agriculture and ecology.'

Organisation of Cells

The content summary of what students are expected to cover in this module are as follows:

- compare the differences between unicellular, colonial and multicellular organisms by:
  - investigating structures at the level of the cell and organelle
  - relating structure of cells and cell specialisation to function
- investigate the structure and function of tissues, organs and systems and relate those functions to cell differentiation and specialisation
- justify the hierarchical structural organisation of organelles, cells, tissues, organs, systems and organisms

Nutrient and Gas Requirements

- investigate the structure of autotrophs through the examination of a variety of materials, for example:
  - dissected plant materials
  - microscopic structures
  - using a range of imaging technologies to determine plant structure
- investigate the function of structures in a plant, including but not limited to:
  - tracing the development and movement of the products of photosynthesis
- investigate the gas exchange structures in animals and plants through the collection of primary and secondary data and information, for example:
  - microscopic structures: alveoli in mammals and leaf structure in plants
  - macroscopic structures: respiratory systems in a range of animals
- interpret a range of secondary-sourced information to evaluate processes, claims and conclusions that have led scientists to develop hypotheses, theories and models about the structure and function of plants, including but not limited to:
  - photosynthesis
  - transpiration-cohesion-tension theory
- trace the digestion of foods in a mammalian digestive system, including:
  - physical digestion
  - chemical digestion
  - absorption of nutrients, minerals and water
  - elimination of solid waste
- compare the nutrient and gas requirements of autotrophs and heterotrophs

Transport

- investigate transport systems in animals and plants by comparing structures and components using physical and digital models, including but not limited to:
  - macroscopic structures in plants and animals
  - microscopic samples of blood, the cardiovascular system and plant vascular systems
- investigate the exchange of gases between the internal and external environments of plants and animals
- compare the structures and function of transport systems in animals and plants, including but not limited to:
  - vascular systems in plants and animals
  - open and closed transport systems in animals
- compare the changes in the composition of the transport medium as it moves around an organism

Module 3 Biological diversity

The NESA Syllabus gives the following content focus for this module: 'Biodiversity is important to balance the Earth's ecosystems. Biodiversity can be affected slowly or quickly over time by natural selective pressures. Human impact can also affect biodiversity over a shorter time period. In this module, students learn about the Theory of Evolution by Natural Selection and the effect of various selective pressures.

Monitoring biodiversity is key to being able to predict future change. Monitoring, including the monitoring of abiotic factors in the environment, enables ecologists to design strategies to reduce the effects of adverse biological change. Students investigate adaptations of organisms that increase the organism's ability to survive in their environment.'

The content summary of what students are expected to cover in this module are as follows:

Effects of the Environment on Organisms

- predict the effects of selection pressures on organisms in ecosystems, including:
  - biotic factors
  - abiotic factors
- investigate changes in a population of organisms due to selection pressures over time, for example:
  - cane toads in Australia
  - prickly pear distribution in Australia

Adaptations

- conduct practical investigations, individually or in teams, or use secondary sources to examine the adaptations of organisms that increase their ability to survive in their environment, including:
  - structural adaptations

- physiological adaptations
- behavioural adaptations

- investigate, through secondary sources, the observations and collection of data that were obtained by Charles Darwin to support the Theory of Evolution by Natural Selection, for example:
  - finches of the Galapagos Islands
  - Australian flora and fauna

Theory of Evolution by Natural Selection

- explain biological diversity in terms of the Theory of Evolution by Natural Selection by examining the changes in and diversification of life since it first appeared on the Earth
- analyse how an accumulation of microevolutionary changes can drive evolutionary changes and speciation over time, for example:
  - evolution of the horse
  - evolution of the platypus
- explain, using examples, how Darwin and Wallace's Theory of Evolution by Natural Selection accounts for:
  - convergent evolution
  - divergent evolution
- explain how punctuated equilibrium is different from the gradual process of natural selection

Evolution – the Evidence

- investigate, using secondary sources, evidence in support of Darwin and Wallace's Theory of Evolution by Natural Selection, including but not limited to:
  - biochemical evidence, comparative anatomy, comparative embryology and biogeography
  - techniques used to date fossils and the evidence produced
- explain modern-day examples that demonstrate evolutionary change, for example:
  - the cane toad
  - antibiotic-resistant strains of bacteria

Module 4 Ecosystem Dynamics

The NESA Syllabus gives the following content focus for this module:
'The Earth's biodiversity has increased since life first appeared on the planet. The Theory of Evolution by Natural Selection can be used to explain periodic increases and decreases in populations and biodiversity. Scientific knowledge derived from the fossil record, and geological evidence has enabled scientists to offer valid explanations for this progression in terms of biotic and abiotic relationships. Students engage in the study of past ecosystems and create models of possible future ecosystems so that human impact on biodiversity can be minimised. The study of ecosystem

dynamics integrates a range of data that can be used to predict environmental change into the future.'

The content summary of what students are expected to cover in this module are as follows:

Population Dynamics

- investigate and determine relationships between biotic and abiotic factors in an ecosystem, including:
  - the impact of abiotic factors
  - the impact of biotic factors, including predation, competition and symbiotic relationships
  - the ecological niches occupied by species
  - predicting consequences for populations in ecosystems due to predation, competition, symbiosis and disease
  - measuring populations of organisms using sampling techniques
- explain a recent extinction event

Past Ecosystems

- analyse palaeontological and geological evidence that can be used to provide evidence for past changes in ecosystems, including but not limited to:
  - Aboriginal rock paintings
  - rock structure and formation
  - ice core drilling
- investigate and analyse past and present technologies that have been used to determine evidence for past changes, for example:
  - radiometric dating
  - gas analysis
- analyse evidence that present-day organisms have evolved from organisms in the past by examining and interpreting a range of secondary sources to evaluate processes, claims and conclusions relating to the evolution of organisms in Australia, for example:
  - small mammals
  - sclerophyll plants
- investigate the reasons for changes in past ecosystems, by:
  - interpreting a range of secondary sources to develop an understanding of the changes in biotic and abiotic factors over short and long periods of time
  - evaluating hypotheses that account for identified trends (ACSBL001)

Future Ecosystems

- investigate changes in past ecosystems that may inform our approach to the management of future ecosystems, including:
  - the role of human-induced selection pressures on the extinction of species

  - models that humans can use to predict future impacts on biodiversity
  - the role of changing climate on ecosystems
- investigate practices used to restore damaged ecosystems, Country or Place, for example:
  - mining sites
  - land degradation from agricultural practices

Full details all of these modules are to be found in the official NESA Biology syllabus (https://syllabus.nesa.nsw.edu.au/assets/global/files/guide-to-the-new-syllabus-in-biology.pdf).

What can I do?

- Become familiar with the course outline early in the first semester to develop the big picture of the whole course.
- Keep up to date with work requirements and any other tasks your teacher sets.
- Review topics regularly.
- When revising, practise questions relevant to a topic, and write notes concentrating on your areas of weakness. No matter how bad your notes, they are the best set of notes for you to develop an increased understanding of a study.
- Sort problems out as they arise by talking to your teacher or brainstorming with your peers.
- Apportion your study effort equally across all your subjects. Within a study, you should spend your study time in proportion to the marks allotted.
- Aim to keep fit and sleep well. A brain well supplied with oxygen works better than one lacking an efficient oxygen supply!

Hints for answering the end-of-year examination

- Highlight the key words in the articles and questions as you answer them – but don't mark the paper during reading time!
- State the obvious. Do what the question asks. When asked a question, answer yes, no, agree, or disagree, then support your answer.
- Keep your answers simple. Point form is often sufficient.
- Do not use biological terms that you do not understand. You will probably not use them in the correct context. Most ideas can be explained by including simple commonly used words.
- Use pen not pencil. Ageing markers have difficulty reading pencil late at night!
- Use a ruler to read graphs accurately. Do not just guess! Near enough may not be good enough!

- Make sure your writing can be read. Scrawling does not hide the fact that you cannot spell or are not sure of the answer.
- Do not spend too long on a question. If you are stuck on a question, move to another question, and return to the 'sticking point' later if you have time.
- Stay relaxed and calm. If you are finding a question difficult, chances are your fellow students will also find the question difficult. A well-reasoned answer will be rewarded.
- Good luck and don't give up!

Chapter 2
Cells as the Basis of Life
Keywords and Terms:

Working Scientifically

accuracy	controlled variable	dependent variable
experimental control group	hypothesis	independent variable
precision	qualitative data	quantitative data
reliability	treatment group	validity

Cell Structure

cell	cell wall	centriole
chloroplast	chromosome	cilia
cytoplasm	DNA	endoplasmic reticulum
eukaryotic	fluid mosaic model	Golgi apparatus
lysosome	magnification	membrane
mitochondrion	nucleolus	nucleus
organelle	permeable	phospholipid bilayer
prokaryotic	prokaryotic	resolution
ribosome	semi-permeable	vacuole
vesicles		

Cell Function

active site	active transport	aerobic respiration
anaerobic respiration	autotroph	catalyst
cellular respiration	chemical energy	concentration gradient
denature	diffusion	endocytosis
enzyme	enzyme-substrate complex	exocytosis
heterotroph	hydrophilic	hydrophobic
hypertonic	hypotonic	induced fit model
inorganic	isotonic	light energy
lock and key model	organic	osmosis
photosynthesis	product	reaction rate
SA:V	substrate	

Summary and Content Review Questions

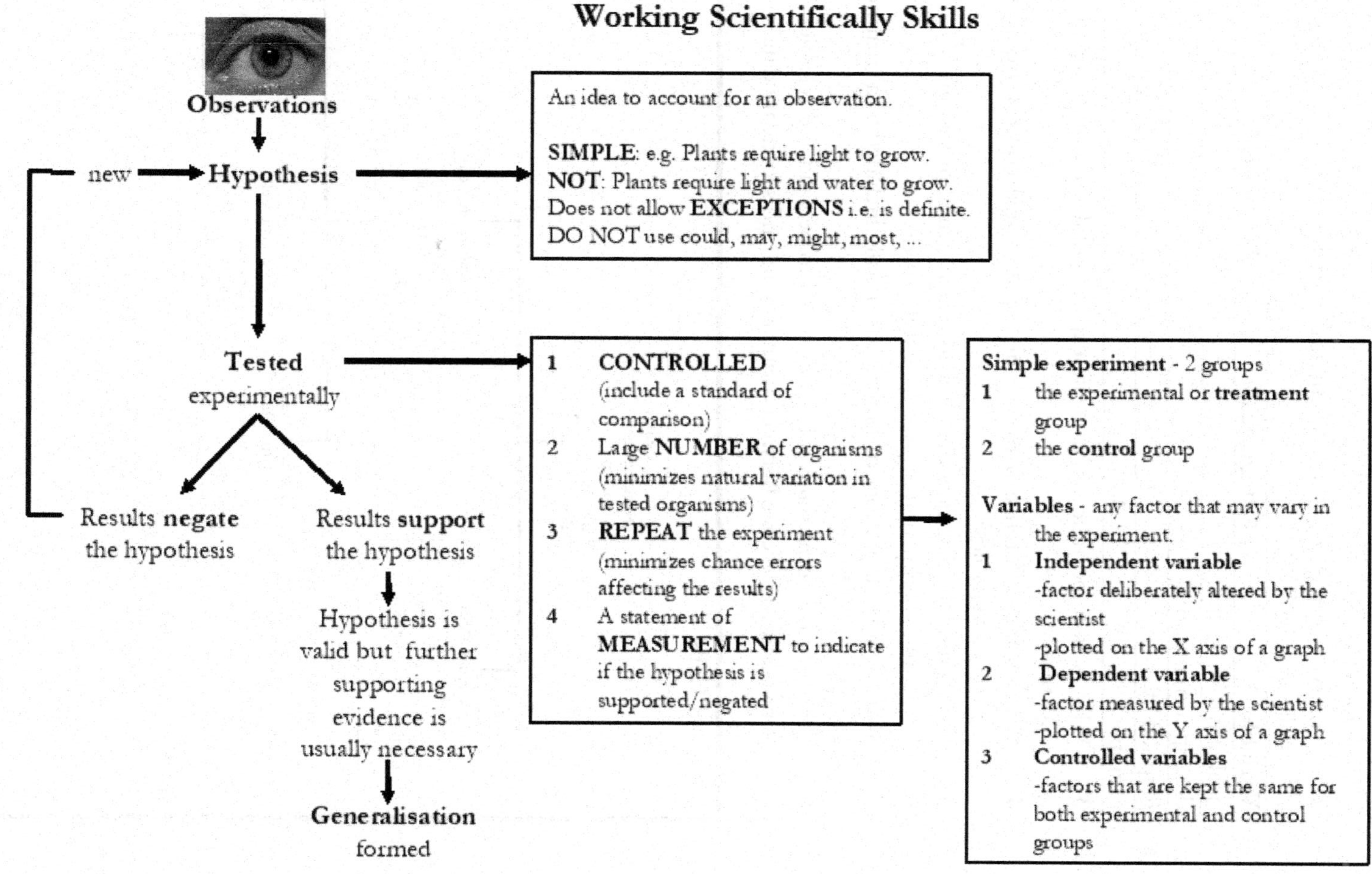

Working scientifically

1 What is meant by 'independent', 'dependent' and 'controlled' variables?

2 Compare categorical and numerical variables by completing the following table.

Variable	Categorical	Numerical
qualitative/quantitative		
description/number		
graphed by		
example		

3 Why is it important to have an experimental control group in an experiment?

4 What is the difference between the experimental control group and controlled variables?

5 Is experimental accuracy the same as experimental precision?

6 How can experimental precision be improved?

7 Why are a large number of individuals usually included in a treatment group?

8 When are experimental results considered to be valid and reliable?

9 What is the difference between primary data and secondary data?

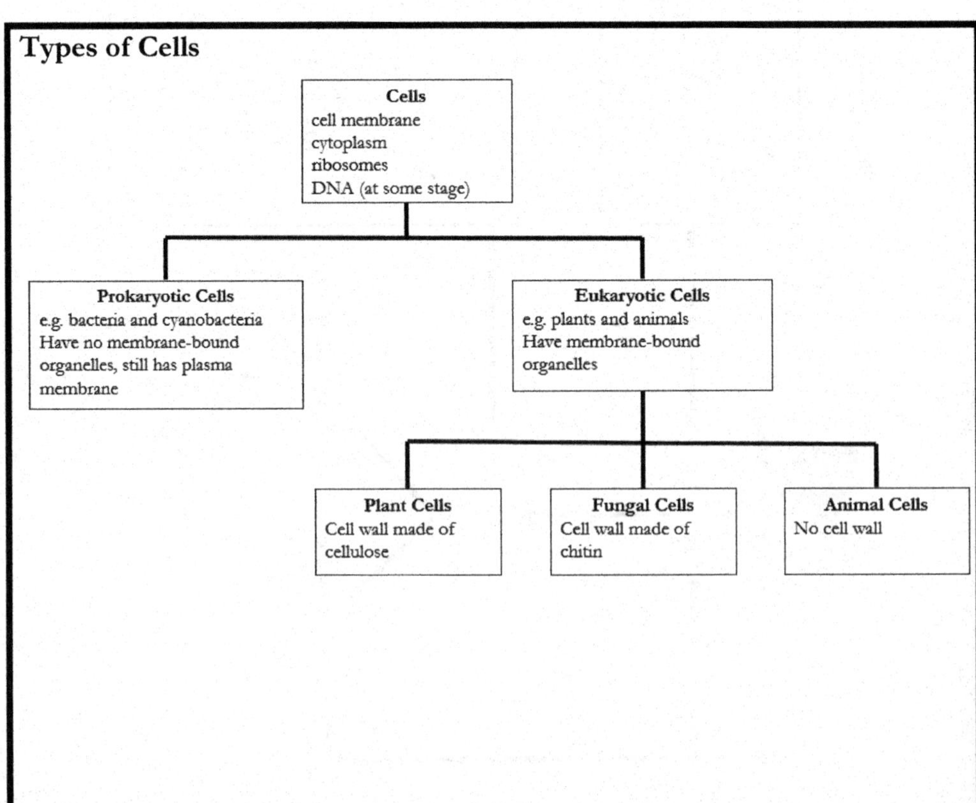

Cell Structure

10 What are the three major points of the 'cell theory'? Name the scientists associated with each point.

11 Draw up a table to compare the human eye, the light microscope and the electron microscope. Consider magnification, resolution, how they work, advantages and disadvantages.

12 What is the difference between resolution and magnification?

13 How is the magnification of an object viewed with a light microscope calculated?

14 Stains are often used for looking at cells with a light microscope. Why are they used?

15 State two disadvantages of the use of stains.

16 Complete the following table summarising the significance of technological advances to developments in the cell theory.

Date	Technological advance	Significance
Mid 1600s	Compound microscope (Robert Hooke)	
Late 1600s	Improved lenses leading to better resolution and magnification (Anton van Leeuwenhoek)	
1800s	Better stains and improved sectioning techniques	
1930s	Phase contrast microscopes (Fritz Zernike)	
1930s	Transmission electron microscopes (Ernst Ruska)	
1930s	Scanning electron microscopes (Max Knoll)	
1930s-40s	Cell fractionation (Albert Claude)	
1940s	Radioactive tracers	
1950s	Confocal microscopes (Marvin Minsky)	

17 Name four features that all cells have in common.

18 Draw up a table to compare the two major types of cells. Include the following: name of each type of cell, their size, structural differences and examples.

19 Which cellular component is found within both eukaryotic and prokaryotic cells?

20 Draw up a table to summarise the function of different parts of a eukaryotic cell. Use the following headings: organelle/cell part, structure, function and observed with light/electron microscope.

21 What would you look for to determine if a cell is prokaryotic or eukaryotic?

22 What would you look for to determine if a cell is from a plant or an animal?

23 How do prokaryotes differ from plant cells?

24 In eukaryotes, photosynthesis occurs in the chloroplast. Draw and describe the structure of the chloroplast. Include in your description where the light dependent and light independent reactions occur.

25 In eukaryotes, aerobic respiration occurs in the mitochondrion. Draw and describe the structure of a mitochondrion.

26 Some scientists think that mitochondria and chloroplasts were once free-living prokaryotes. What structural evidence supports this idea?

27 List 6 functions of membranes in general.

28 Membranes are selectively (partially or semi) permeable. What does this mean?

29 Use the following diagram of the fluid mosaic model of the cell membrane to complete the table that follows.

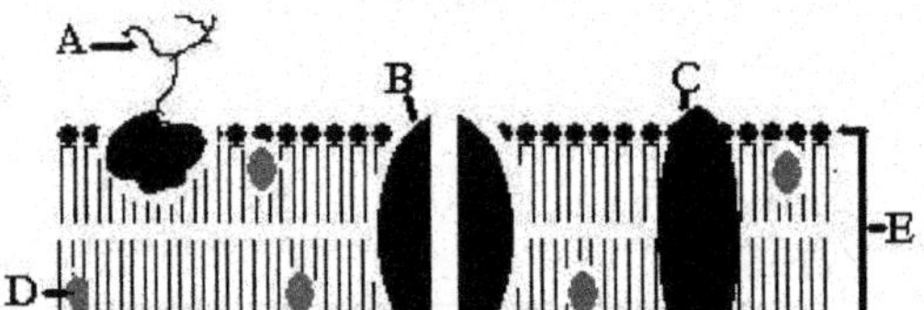

	Structure	Function
A		
B		
C		
D		
E		

30 What are the two major compounds found in membranes?

31 The fluid mosaic model is often used to describe the structure of cell membranes. Describe this model.

Cell Function

32 List four factors that increase the rate of diffusion.

33 What is meant by 'diffusion gradient'?

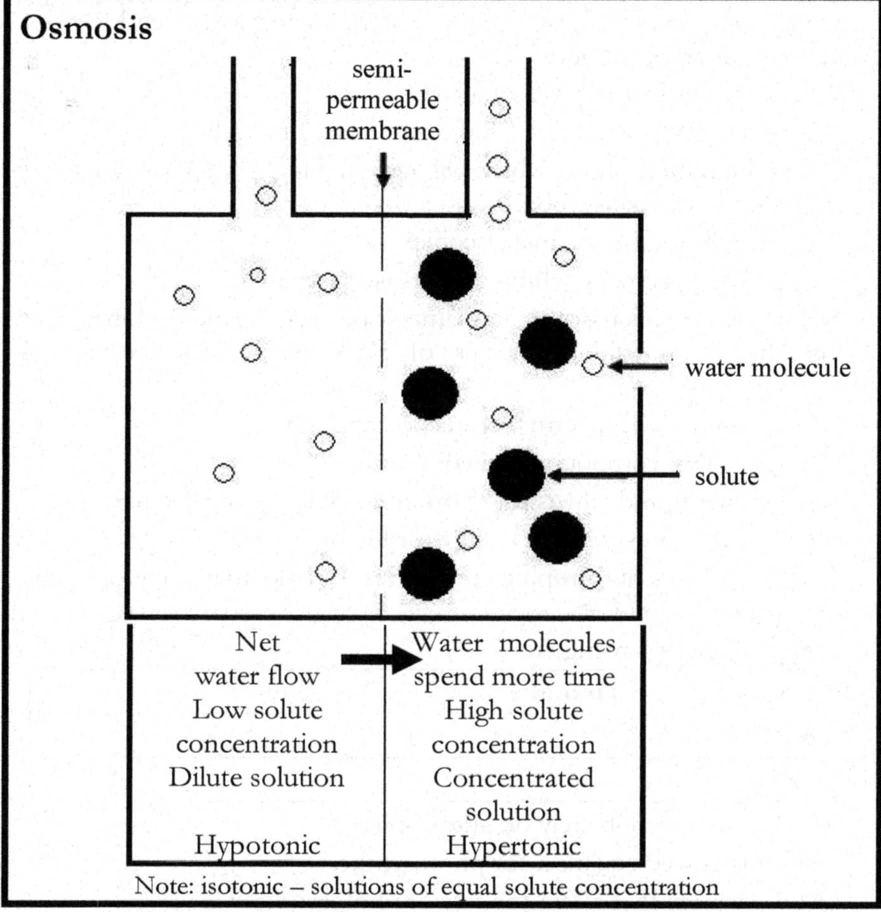

34 Explain why an animal cell placed in distilled water will burst whereas a plant cell placed in distilled water will not burst. In your answer show that you understand what is meant by osmosis.

35 Complete the following summary of movement across membranes.

<table>
<tr><th rowspan="2"></th><th rowspan="2">Simple Diffusion</th><th rowspan="2">Channel-mediated</th><th colspan="2">Carrier-mediated</th><th rowspan="2">Vesicle-mediated (exocytosis/ endocytosis)</th></tr>
<tr><th>Facilitated diffusion</th><th>Active transport</th></tr>
<tr><td>Where</td><td></td><td></td><td colspan="2"></td><td></td></tr>
<tr><td>What</td><td></td><td></td><td></td><td></td><td></td></tr>
<tr><td>Direction</td><td colspan="3"></td><td></td><td></td></tr>
<tr><td>Energy</td><td colspan="3"></td><td colspan="2"></td></tr>
</table>

36 Define and compare the processes: diffusion, osmosis, facilitated diffusion, active transport and vesicle mediated transport.

37 In summary, what factors determine where and how hydrophilic and hydrophobic substances cross membranes?

38 How are the following terms related: exocytosis, endocytosis, pinocytosis and phagocytosis?

39 What is meant by the SA:V of a cell?

40 How is the SA:V of a cell related to the size of the cell?

41 Given the same volume, which will have the highest SA:V – a cube, a sphere or a rectangular box?

42 Why do cells tend to be microscopic?

43 What features do cells exhibit to increase their SA:V?

44 Nerve cells are microscopic in diameter but may be metres long. How can this be possible in terms of SA:V and a cell meeting its requirements?

45 What is meant by the term 'organic' compound?

46 What are other compounds called?

47 What is energy and why is it important to all living organisms?

48 What is the main source of energy for life on Earth?

49 What is it that 'autotrophic' organisms can do that 'heterotrophic' organisms cannot do?

50 Complete the following table.

Plants use	**To make**
Nitrates	
Phosphates	

51 Why do heterotrophs rely on autotrophs?

52 Write a balanced equation for photosynthesis.

53 List the factors that affect the rate of photosynthesis.

54 Write a balanced chemical equation for aerobic respiration.

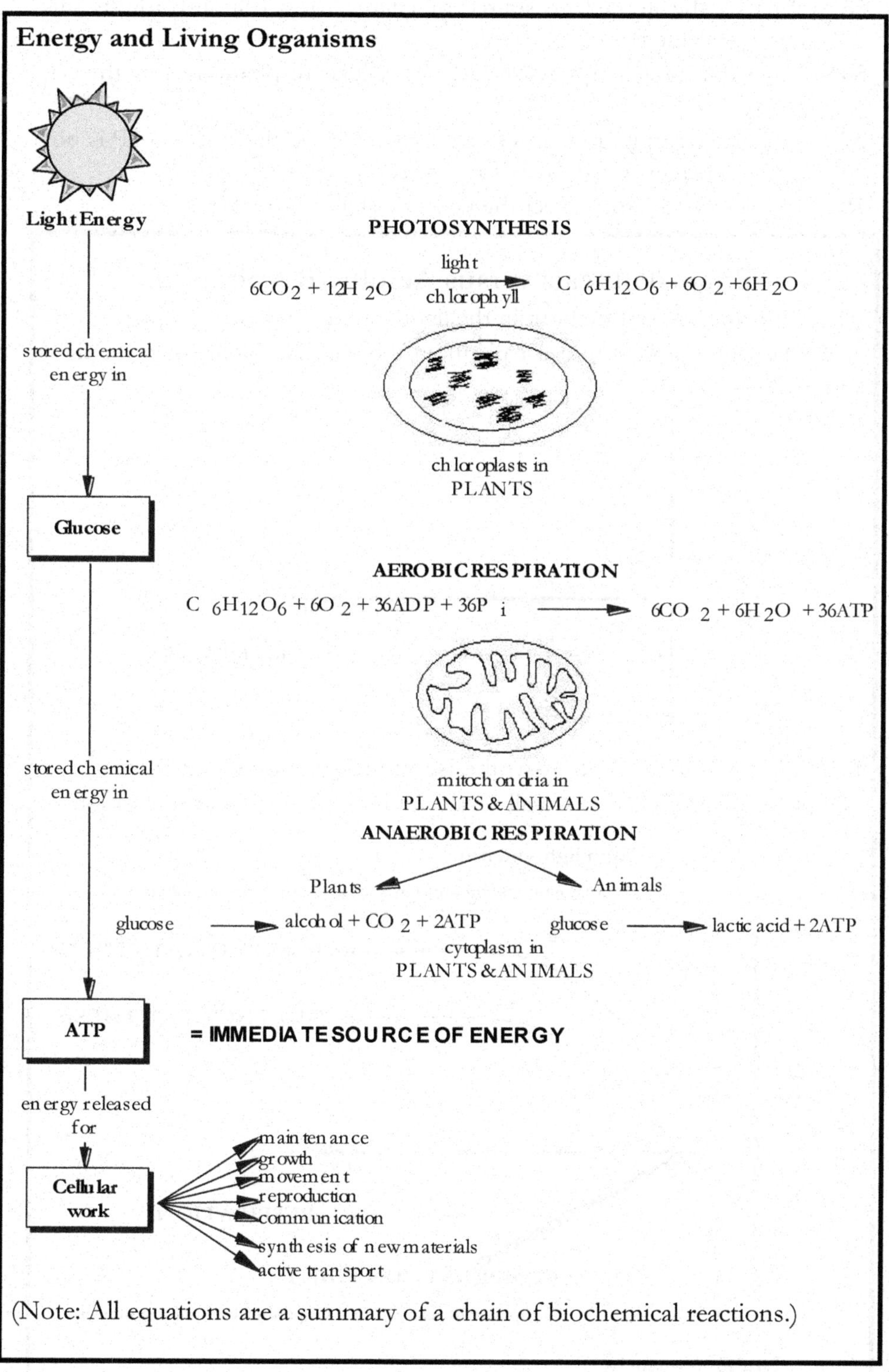

(Note: All equations are a summary of a chain of biochemical reactions.)

55 What is the immediate source of energy for a cell and how is this energy released?

56 What does the energy released from cellular respiration allow the cell to do?

57 In aerobic respiration the energy transfer from glucose to ATP is not 100% efficient. What other form of energy is produced?

58 List the factors that affect the rate of cellular respiration.

Photosynthesis versus Aerobic Respiration

Plants and animals respire both in the light and in the dark.

The following graph is true for plants in the dark and animals in the light and dark.

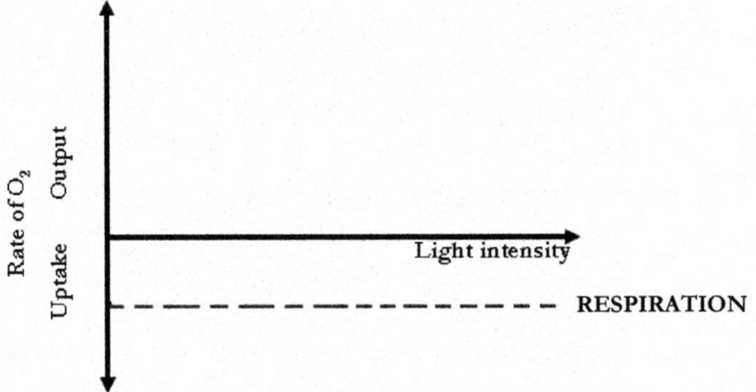

In the light, plants both photosynthesise and respire at the same time.

Net photosynthesis is the overall result of photosynthesis and respiration.

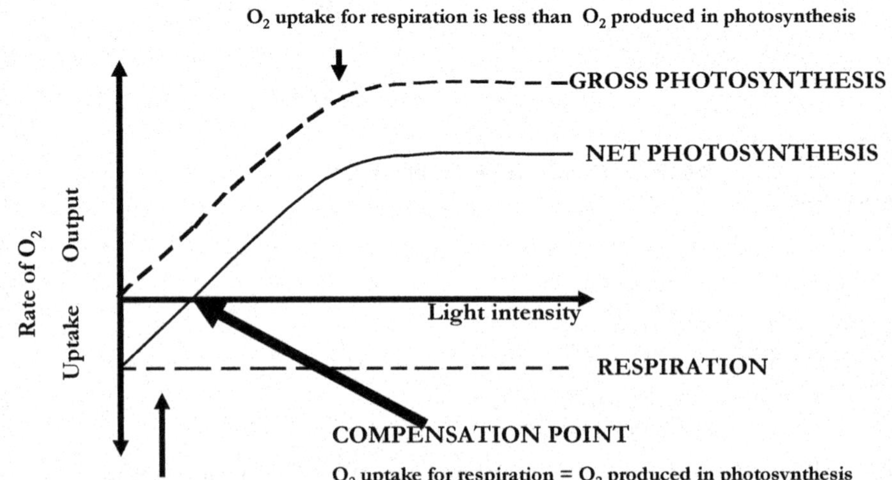

59 Complete the following table.

Biochemical process	Reactants	Products
Photosynthesis		
Aerobic respiration		
Anaerobic respiration in plants		
Anaerobic respiration in animals		

60 What are three differences between aerobic respiration and anaerobic respiration?

61 Use your knowledge of photosynthesis and respiration to complete the following comparison of autotrophs and heterotrophs. Consider whether CO_2, O_2, H_2O, N, P and S are required or produced.

	Autotrophs	Heterotrophs
Energy source		
CO_2		
O_2		
H_2O		
N		
P		
S		

62 Why can rate of oxygen uptake and output be used to measure overall rate of photosynthesis?

63 What else could be used to measure overall rate of photosynthesis?

64 Complete the following table.

	Material	Where they cross the cell membrane	Process
	Oxygen		
	Water		
	Glucose		
	Carbon dioxide		
	Urea		

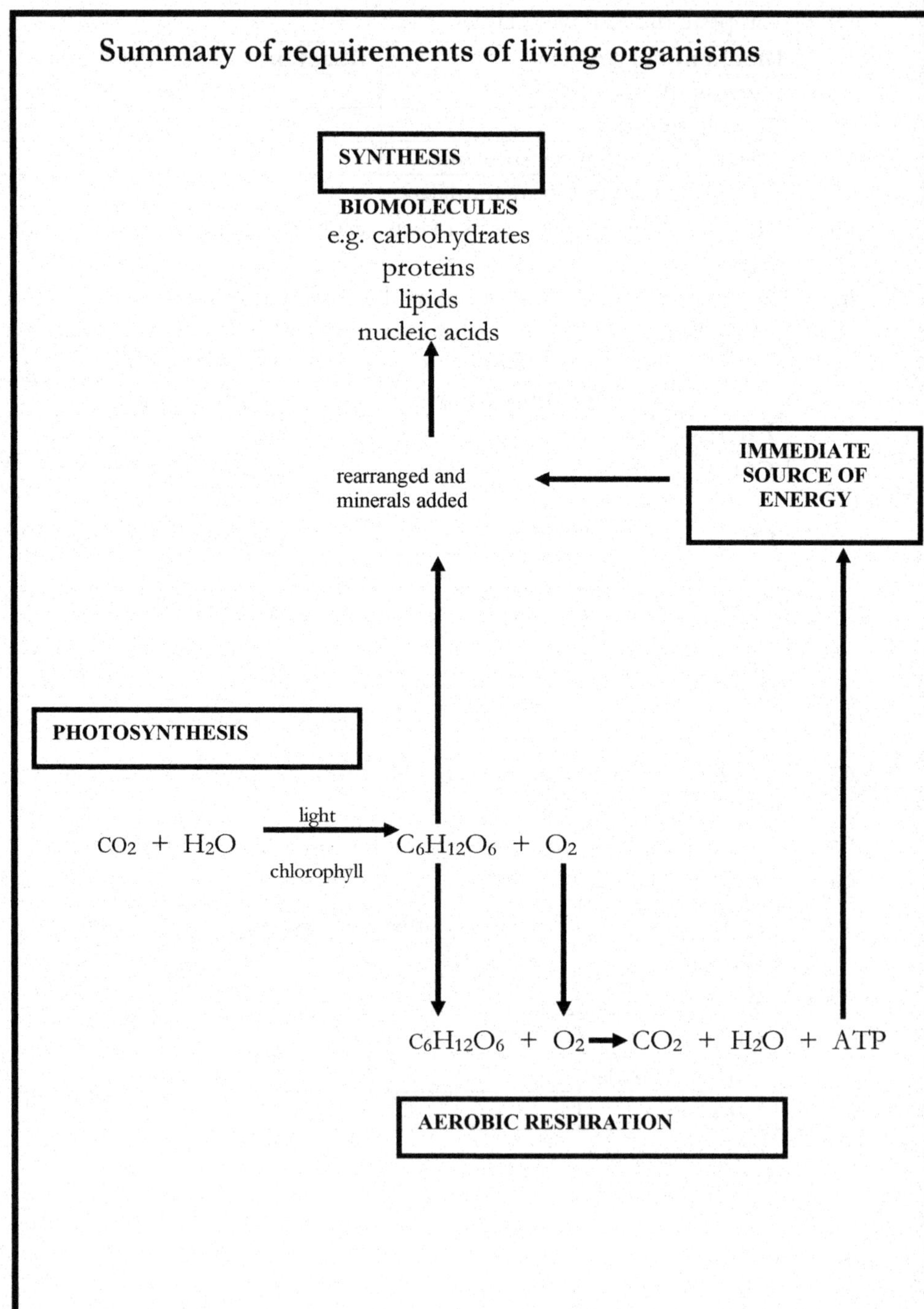
Summary of requirements of living organisms
SYNTHESIS
BIOMOLECULES
e.g. carbohydrates
proteins
lipids
nucleic acids
rearranged and
minerals added
IMMEDIATE
SOURCE OF
ENERGY
PHOTOSYNTHESIS
CO_2 + H_2O
light
chlorophyll
$C_6H_{12}O_6$ + O_2
$C_6H_{12}O_6$ + O_2 → CO_2 + H_2O + ATP
AEROBIC RESPIRATION

Enzymes

-increase the rate of reactions

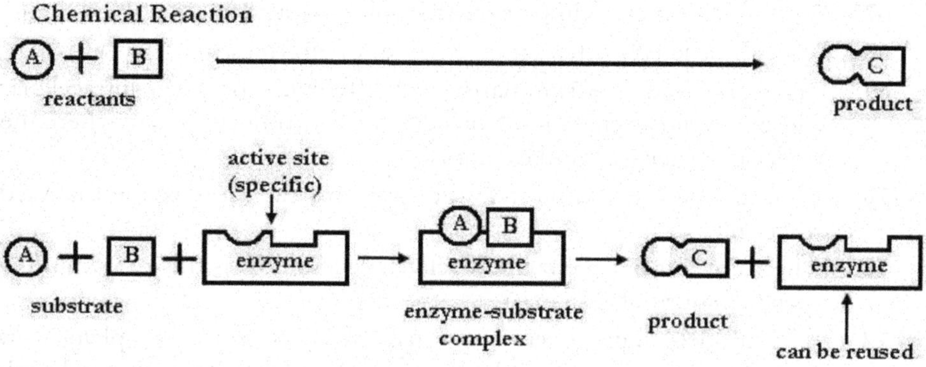

Chemical Reaction

A + B → C

reactants

product

active site (specific)

A + B + enzyme → enzyme (A B) → C + enzyme

substrate

enzyme-substrate complex

product

can be reused

Enzymes and reaction rates

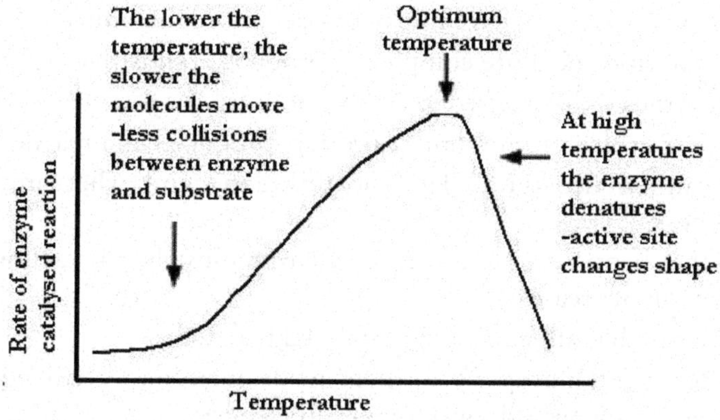

The lower the temperature, the slower the molecules move -less collisions between enzyme and substrate

Optimum temperature

At high temperatures the enzyme denatures -active site changes shape

Rate of enzyme catalysed reaction

Temperature

If rate of enzyme catalysed reaction is plotted against pH, the curve has a similar shape.

Optimum pH and temperature are usually close to the pH and temperature of the organism in which the enzyme is found.

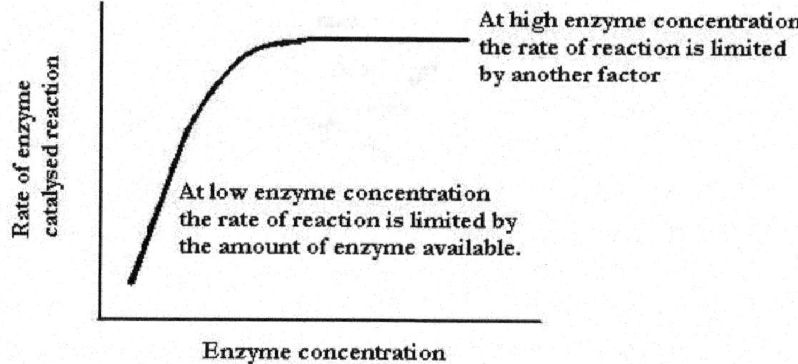

If the rate of enzyme catalysed reaction is plotted against substrate concentration, the curve has a similar shape.

65 Why are enzymes sometimes called organic catalysts?

66 What is meant by 'rate of reaction'?

67 Why are enzymes so important to living organisms?

68 What three letters often end the names of enzymes?

69 Draw up a table to summarise the differences between intracellular and extracellular enzymes. Consider site of production, site where the reaction is catalysed and examples.

70 What is another name for the reactants in an enzyme-catalysed reaction?

71 What is meant by an enzyme's active site and how is this related to the specific nature of enzymes?

72 Two models have been suggested by scientists to explain how enzymes work. These are the 'lock and key' model and the 'induced-fit' model. How do these models differ?

73 Arrange the following into an equation that shows how enzymes speed up reactions. Note, a term may be used more than once.

enzyme enzyme-substrate complex product substrate

74 Why are enzymes only required in small amounts?

75 Do enzymes increase the amount of product produced in a reaction?

76 Enzymes can be denatured. How is this done and what are the consequences?

77 Describe what happens to the rate of an enzyme-catalysed reaction as the temperature increases.

78 List five factors that affect enzyme-catalysed reactions.

79 Mitochondria and chloroplasts contain many membranes embedded with enzymes. How would this increase efficiency?

Section I Questions

80 Below is a diagram of a field of view from a light microscope. The divisions on the scale are 30 μm apart.

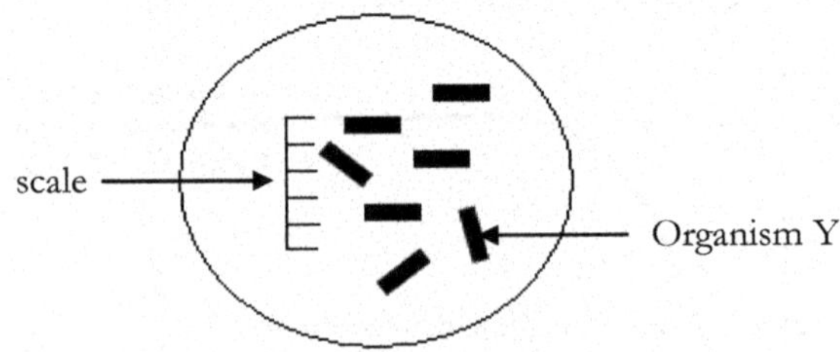

The length of Organism Y is

(A) 15 μm.
(B) 30 μm.
(C) 60 μm.
(D) 300 μm.

81 All cells have
(A) nuclei.
(B) vacuoles.
(C) ribosomes.
(D) mitochondria.

82 Which of the following organelles would be found in the cells of a prokaryote?
(A) ribosomes
(B) chloroplasts
(C) vacuoles
(D) a nucleus

83 Prokaryotic cells
(A) cannot photosynthesise because they do not have chloroplasts.
(B) are larger than eukaryotic cells.
(C) contain circular chromosomes.
(D) only reproduce asexually.

84 All bacteria are
(A) harmful.
(B) decomposers.
(C) prokaryotes.
(D) photosynthetic.

85 In eukaryotic cells
(A) the rough endoplasmic reticulum is the site of protein synthesis.
(B) the Golgi body is the site of aerobic respiration.
(C) the mitochondrion is the site of lipid synthesis.
(D) lysosomes store pigments.

86 Four different, appropriately stained cells were viewed with a light microscope. The cells included a muscle cell, a pancreatic cell, a beetroot cell and a photosynthesising cell. For each cell, it would be reasonable to expect the student to be able to see
(A) chromosomes.
(B) chloroplasts.
(C) vacuoles.
(D) a nucleus.

87 A structure not found in all living plant cells is a
(A) mitochondrion.
(B) ribosome.
(C) chloroplast.
(D) cell wall.

88 In plant and animal cells

(A) the rough endoplasmic reticulum is the site of protein synthesis.
(B) the Golgi body is the site of aerobic respiration.
(C) the mitochondrion is the site of lipid synthesis.
(D) lysosomes store pigments.

89 Membranes

(A) control what enters and leaves a cell.
(B) are composed of a protein bilayer.
(C) are impermeable.
(D) are rigid.

Use the following information to answer Questions 90 and 91.

An artificial cell containing only water; is suspended in a sucrose solution. The membrane of the artificial cell is permeable to water but not to sucrose.

90 After 4 hours

(A) the artificial cell will have exploded.
(B) the artificial cell will have increased in size.
(C) the artificial cell will have decreased in size.
(D) sucrose will have moved into the artificial cell.

91 The changes observed are due to

(A) diffusion.
(B) facilitated diffusion.
(C) osmosis.
(D) active transport.

92 Sodium ions may be taken into a cell by active transport. Active transport

(A) occurs through protein channels in the membrane.
(B) occurs through protein carrier molecules in the membrane.
(C) does not require energy from the cell.
(D) involves movement down a concentration gradient.

93 When comparing diffusion and active transport:

(A) in diffusion, molecules move from where they are in low concentration to where they are in high concentration whereas in active transport the molecules move from where they are in high concentration to where they are in low concentration.
(B) in both, molecules move from where they are in high concentration to where they are in low concentration.
(C) in both, molecules move from where they are in low concentration to where they are in high concentration.
(D) in active transport, molecules move from where they are in low concentration to where they are in high concentration whereas in diffusion the molecules move from where they are in high concentration to where they are in low concentration.

94 Cells are small because this provides

(A) a small surface area compared with volume for exchange of molecules with the environment.
(B) a large surface area compared with volume for exchange of molecules with the environment.
(C) a large volume compared with surface area for exchange of molecules with the environment.
(D) a large surface area and volume for exchange of molecules with the environment.

Use the following information to answer Questions 95 and 96.

The following cells have the same volume.

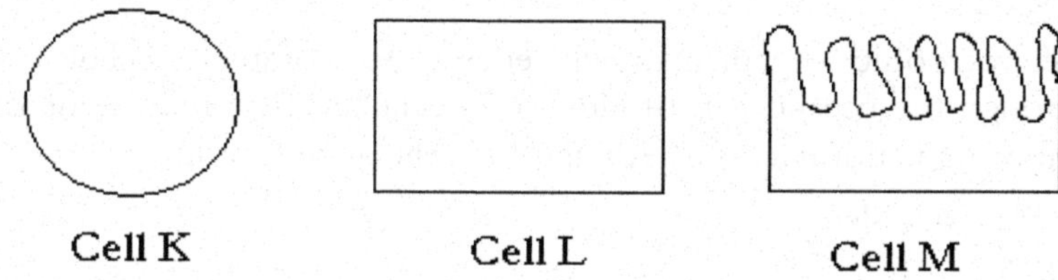

95 Which of the cells has the lowest surface area to volume ratio?

(A) cell K
(B) cell L
(C) cell M
(D) all have the same SA:V

96 Cell M has many mitochondria. It is likely that cell M

(A) is a site of active uptake of mineral ions.
(B) exchanges materials with the surroundings by diffusion.
(C) is a storage cell.
(D) is involved with transport of oxygen.

97 With respect to organic molecules:

(A) they contain carbon.
(B) they cannot be broken down.
(C) water is an example.
(D) they are only made by plants.

98 Organic molecules contain

(A) carbon, hydrogen and ozone.
(B) carbon, hydrogen and oxygen.
(C) carbon, nitrogen and iron.
(D) hydrogen and oxygen.

99 Inputs of photosynthesis are

(A) oxygen, water and light.
(B) glucose and carbon dioxide.
(C) carbon dioxide, water and ATP.
(D) carbon dioxide, water and light.

100 When green plants photosynthesise, they produce complex organic substances. In this process,

(A) light energy is transformed into and stored as heat energy.
(B) light energy is converted into and stored as chemical energy.
(C) chemical energy is transformed into and stored as light energy.
(D) heat energy is converted into and stored as chemical energy.

101 In photosynthesis, producers convert light energy to chemical energy. One of the following is not directly necessary for photosynthesis.

(A) carbon dioxide
(B) water
(C) chlorophyll
(D) oxygen

102 Heterotrophic organisms gain energy to complete cellular work through the breakdown of glucose in cellular respiration. One of the following is not necessary for aerobic cellular respiration

(A) carbon dioxide.
(B) water.
(C) glucose.
(D) oxygen.

103 Both plants and animals require inputs of

(A) oxygen.
(B) carbon dioxide.
(C) light.
(D) protein.

104 Phototrophs

(A) never take up oxygen from their surroundings as they produce oxygen in photosynthesis.
(B) convert the energy in sunlight into the energy stored in the bonds of organic molecules.
(C) never release carbon dioxide to the atmosphere as carbon dioxide is used in photosynthesis.
(D) convert the energy released from simple inorganic reactions into the energy stored in the bonds of organic molecules.

105 Light energy is the energy source for most ecosystems. Light energy is used by autotrophs to produce organic material. The energy has been converted to chemical energy that can be stored. When compounds are broken down, some energy is unavoidably lost as

(A) light energy.
(B) chemical energy.
(C) heat energy.
(D) kinetic energy.

106 Enzymes

(A) increase the speed of reactions that would not occur without the enzyme.
(B) are permanently changed by the reaction.
(C) are steroids
(D) are reused.

107 Pepsin is a protease that is secreted into the human stomach. Pepsin would

(A) have an optimum pH of 2.
(B) catalyse the breakdown of lipids.
(C) have an optimum temperature of 22°C.
(D) would be destroyed by the acid in the gastric juices.

Section II Questions

108

The diagram below is of a cell seen under an electron microscope.

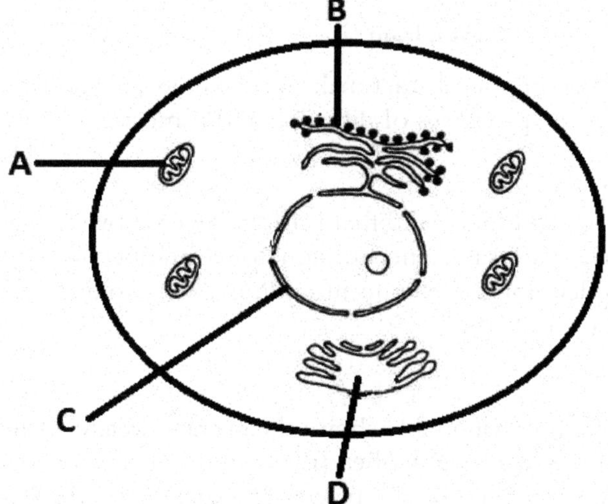

(a) Is the cell a plant or animal cell? Explain your answer. [2 marks]

(b) Complete the following table for structures labelled in the diagram.

	Structure	Function
A		major site of ATP production
B	ribosome	
C	nuclear membrane	
D		packages material for secretion out of the cell

[4 marks]

(c) Why are stains often used when viewing sections with a light microscope? [2 marks]

[Total 8 marks]

109

(a) Research and then assess the impact of improvements in microscopy on our understanding of chloroplasts. [7 marks]

(b) Describe how you gathered and analysed the information in part **(a)**. [3 marks]

(c) How did you evaluate the relevance and reliability of the information gathered in part **(a)**? [2 marks]

[Total 12 marks]

110

The diagram below shows a diagram of a cell membrane.

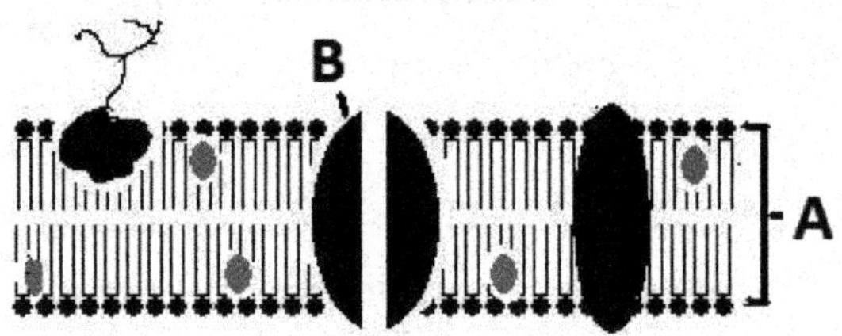

(a) Name the compounds that make up structures A and B. [2 marks]

(b) Compare the processes of diffusion and osmosis. [4 marks]

[Total 6 marks]

111

Sand worms live in beach sand that is often covered by sea water. They can tolerate extremes of salt concentration for a number of hours. Design a procedure to determine the concentration of saline solution that is isotonic to the sand worm's tissue. [4 marks]

[Total 4 marks]

112

The data in the following table shows the average weight of hens' eggs as a percentage of original weight when placed in distilled water. The outer shell of the eggs was removed before placing the eggs in the distilled water. This left the cell membrane undamaged and the cell contents intact.

Time (minutes)	0	10	15	20	25	30
Weight percentage of original	100	103	106	108	111	114

(a) Use the data to draw an appropriate graph. [4 marks]

(b) From the graph, describe how the weight of the eggs changed during the experiment. [1 mark]

(c) Using your knowledge of osmosis, explain the results of the experiment. [2 marks]

[Total 7 marks]

113

The following experiment was set up.

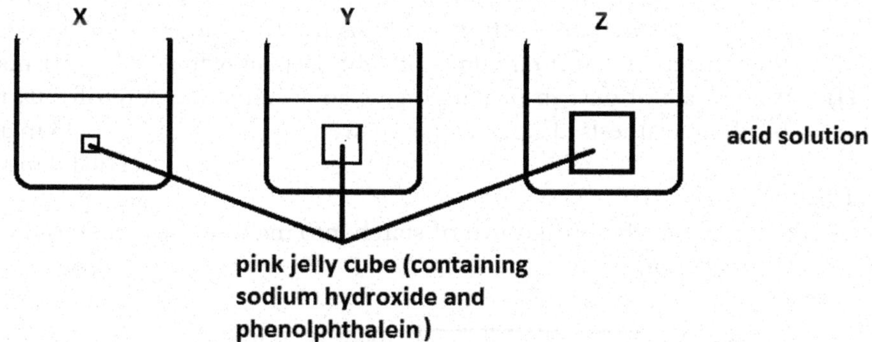

Pink jelly cubes of different sizes but all containing sodium hydroxide and phenolphthalein were placed in the same volume and concentration of acid solution. The acid solution moved into the cubes causing a chemical reaction that changed the colour of the jelly from pink to clear.

(a) What is the name of the process where the acid moved into the jelly? [1 mark]

(b) Which cube became clear first? [1 mark]

(c) Explain your answer to part **(b)**. [1 mark]

[Total 3 marks]

114

Green leaves carry out photosynthesis.

(a) What organelle in green leaf cells is the site of photosynthesis? [1 mark]

(b) What are the inputs and outputs of photosynthesis? [2 marks]

A scientist measured the rate of photosynthesis as the concentration of carbon dioxide in the surroundings of a potato plant, increased. This experiment was repeated at three different light intensities. The results are shown in the graph that follows.

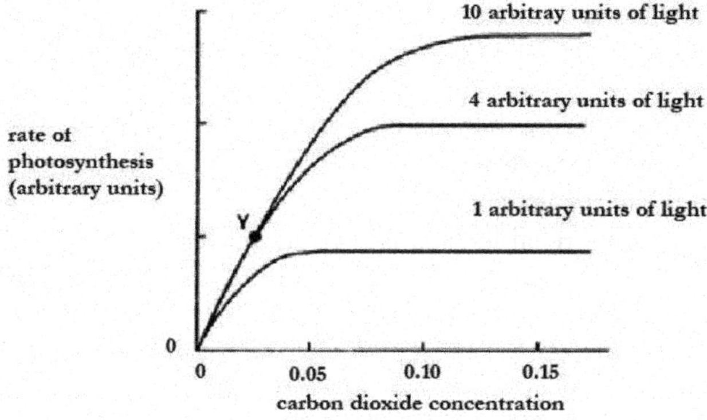

(c) List a factor that should be the same at each light intensity. [1 mark]

(d) Which light intensity resulted in the highest rate of photosynthesis at 0.05 carbon dioxide concentration? [1 mark]

(e) At Point Y, the rate of photosynthesis for 10 arbitrary units of light is the same as at 4 arbitrary units of light. Explain why. [1 mark]

(f) Would rate of oxygen produced give an accurate measure of total rate of photosynthesis? Explain your answer. [3 marks]

[Total 9 marks]

115

Amylase catalyses the breakdown of starch into maltose according to the following equation:

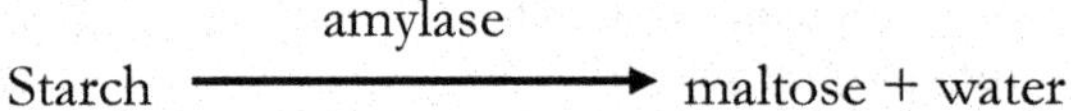

(a) Amylase is an 'organic catalyst'. What does this mean? [1 mark]

A student performed a single experiment using amylase and obtained the following results.

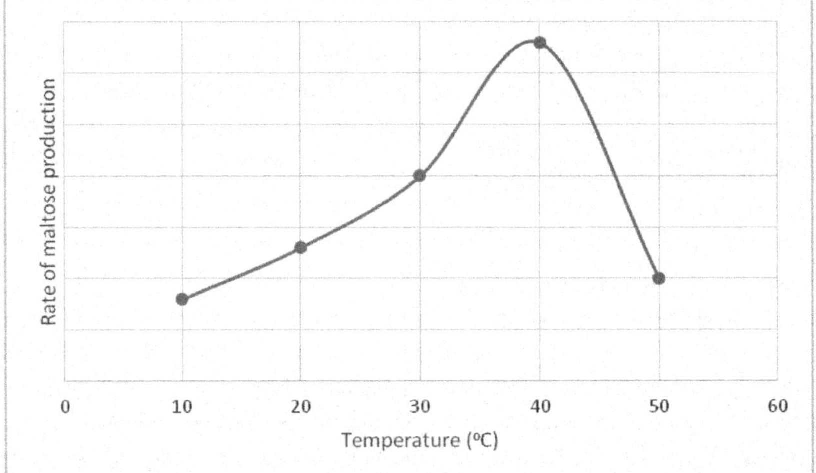

(b) What is the purpose of this experiment? [1 mark]

(c) What is the independent variable and dependent variable in this experiment? [2 marks]

(d) Name one variable that would have been controlled in this experiment. [1 mark]

(e) Describe one change that would improve the validity of the data collected in the experiment. [1 mark]

(f) What do these results suggest? [1 mark]

(g) Are these results reliable? Explain your answer. [1 mark]

[Total 8 marks]

Chapter 3
Organisation of Living Things
Keywords and Terms:

Organisation of Cells

colonial	differentiation	external environment
extracellular	function	internal environment
intracellular	metabolism	multicellular
organ	specialisation	structure
system	tissue	unicellular

Nutrient and Gas Requirements

absorption	alveoli	bile
caecum	carbohydrate	carnivore
chemical digestion	digestion	elimination
excretion	herbivore	ingestion
leaf	lenticel	lipid
physical digestion	protein	root hair
stomata	stomatal aperture	transpiration

Transport

artery	atria	blood
capillary	cardiovascular	closed circulatory system
haemoglobin	open circulatory system	phloem
translocation	vein	ventilation
ventricle	xylem	

Summary and Content Review Questions
Organisation of Cells

1 What is the main advantage of a large size to an organism?

2 Why are large organisms made of many cells instead of one very large cell?

3 Explain the difference between a specialised cell and a differentiated cell.

4 How are cells, tissues, organs and systems related?

5 What is the main problem faced by multicellular organisms as a result of specialisation and differentiation?

6 Complete the following table comparing unicellular, colonial and multicellular organisms.

	Cells within the following organisms		
	Unicellular	**Colonial**	**Multicellular**
Example			
Number of cells			

Ability to undergo cellular respiration			
Degree of specialisation and differentiation			
Ability to perform all functions in order to sustain life			
Ability to survive alone			

7 How do the internal and external environments of unicellular organism differ from those of a multicellular organism?

8 Complete the following table.

Cell	**Function**	**Observation**	**Explanation**
Palisade cell	Photosynthesis	Many chloroplasts present	
Pancreatic cell	Secretion of insulin	Many Golgi bodies and mitochondria	
Small intestinal	Absorption of products of digestion	Presence of microvilli	

9 Two flowering plant systems are the shoot and root systems. Complete the following table summarising these two systems.

Flowering Plant System	**Organs**	**Function**
Shoot system		
Root system		

Nutrient and Gas Requirements

10 What determines the amount and type of nutrients required by an organism?

11 What are the features of effective gas exchange surfaces?

12 Explain how roots help terrestrial plants to photosynthesise.

13 What structural features allow roots to be surfaces of exchange?

14 Using the definition of osmosis, describe how water is taken up by the root.

15 Mineral ions are also taken into the plant by the roots. How does this happen?

16 What are the roles of root hairs and lenticels in gas exchange in plants?

17 Leaves are the major site of photosynthesis. Explain how leaves are structurally adapted for photosynthesis.

18 Describe the structure of stomata.

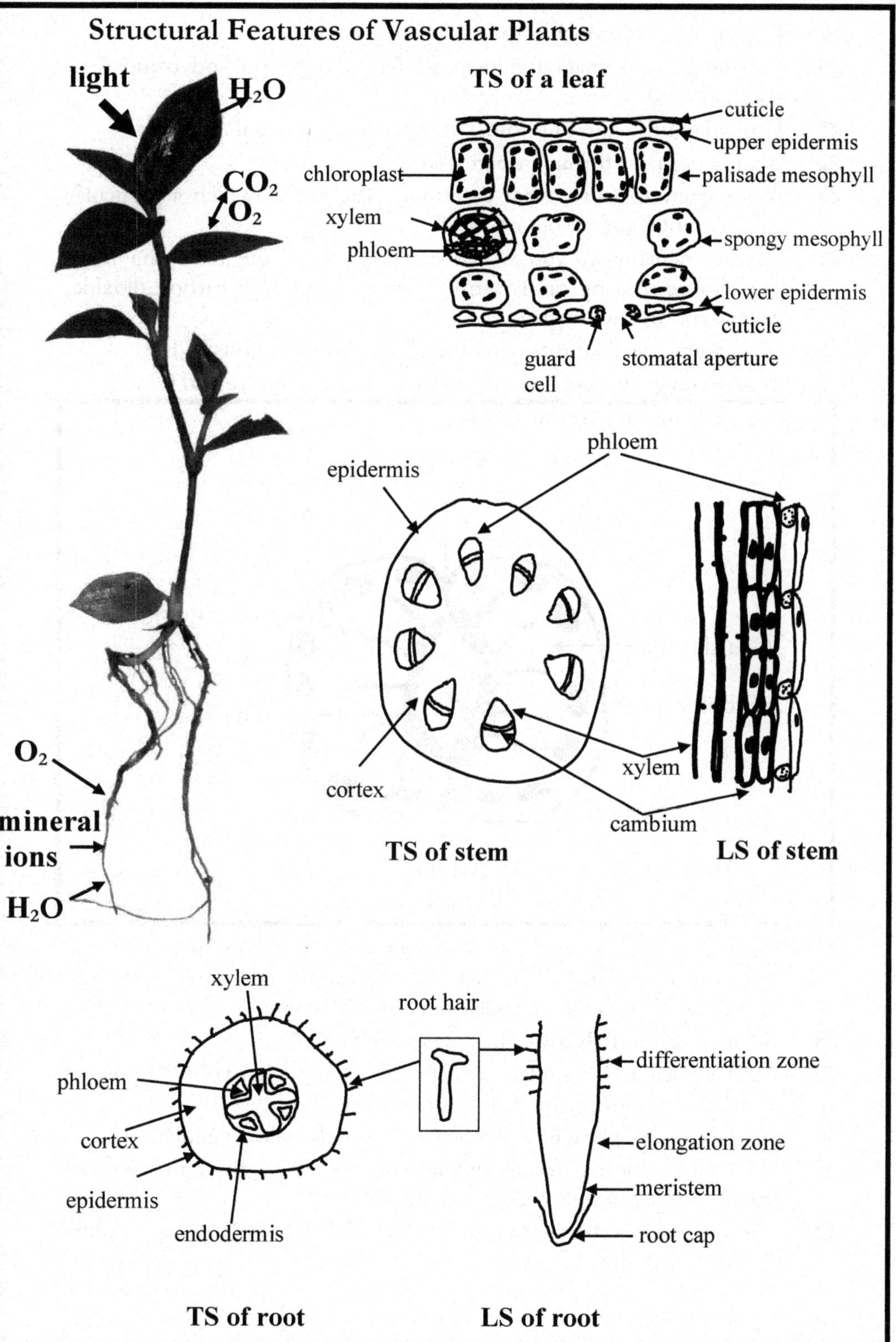
Structural Features of Vascular Plants
light
H_2O
CO_2
O_2
O_2
mineral ions
H_2O
TS of a leaf
cuticle
upper epidermis
chloroplast
palisade mesophyll
xylem
phloem
spongy mesophyll
lower epidermis
cuticle
guard cell
stomatal aperture
phloem
epidermis
xylem
cortex
cambium
TS of stem
LS of stem
xylem
root hair
phloem
cortex
epidermis
endodermis
differentiation zone
elongation zone
meristem
root cap
TS of root
LS of root

19 Do guard cells have chloroplasts? Is this significant?

20 Explain how stomata function. (Refer to diffusion and osmosis in your answer.)

21 How might the rate of photosynthesis effect stomatal aperture?

22 What gases pass through open stomata?

23 Apart from aperture, what determines the rate at which a particular gas will move into or out of a leaf?

24 Explain which way you would expect each of the following three gases to be moving on a warm sunny day and why: carbon dioxide, water vapour, and oxygen.

25 Most plants need stomata open to allow photosynthesis. Why?

26 At what times do most plants reduce stomatal aperture and why?

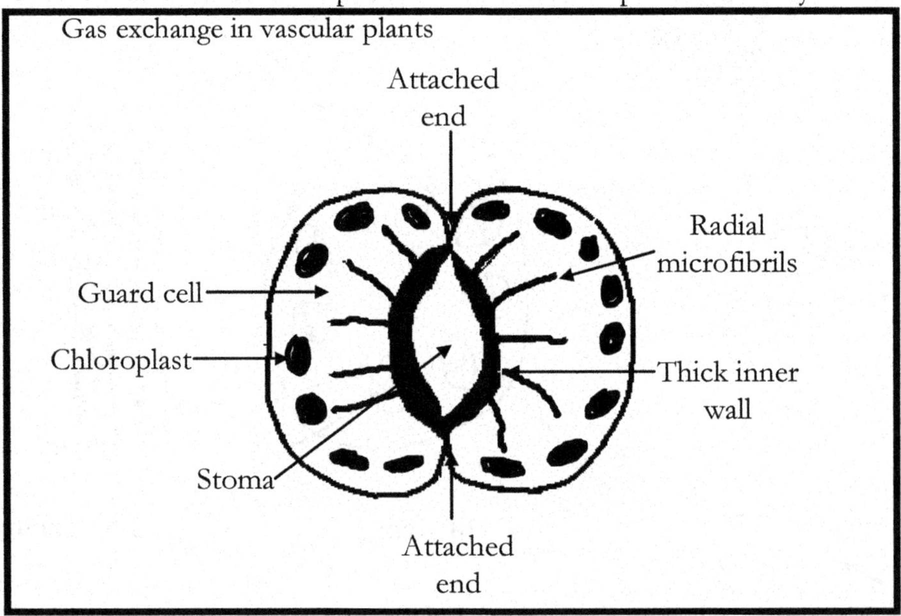

27 Outline the structure and function of the major parts of the mammalian respiratory system in a table (nose, mouth, trachea, bronchus, bronchioles, alveolar ducts and alveoli).

28 What features of alveoli enhance gas exchange?

29 What is meant by ventilation of the lungs and what is its purpose in terms of the some of the features described above.

30 How do the diaphragm and intercostal muscles allow ventilation?

31 How does blood entering and leaving the lungs differ in terms of oxygen and carbon dioxide concentrations?

32 Compare gas exchange in insects, fish and mammals by completing the following table.

	Insect	Fish	Frog	Mammal
Diagram				
Description				
Structure where exchange occurs				
How the surface is kept moist				
Ventilation				
How is the diffusion gradient maintained?				

33 Complete the following comparison of gas exchange in plants and animals.

		Plants	Mammals
Gas Exchange	**Role**		
	Gases		
	Structures		
	Process		

34 Why must food be digested before it can be absorbed into the body?

35 What characteristics do effective digestive systems share?

36 Use the digestion of a starch biscuit to explain the difference between physical and chemical digestion.

37 What is the difference between intracellular and extracellular digestion?

38 Teeth are involved in physical digestion. Why does physical digestion usually occur before physical digestion?

39 Use the following table to summarise digestion in humans.

Structure (include a list of special features)	Major functions	Secretions	Mechanical digestion	Chemical digestion
Mouth				
Oesophagus				
Stomach				
Small intestine				
Large intestine				
Rectum and anus				

40 Why is bile considered to be involved in physical digestion?

41 Would salivary amylase catalyse the breakdown of carbohydrates in the stomach?

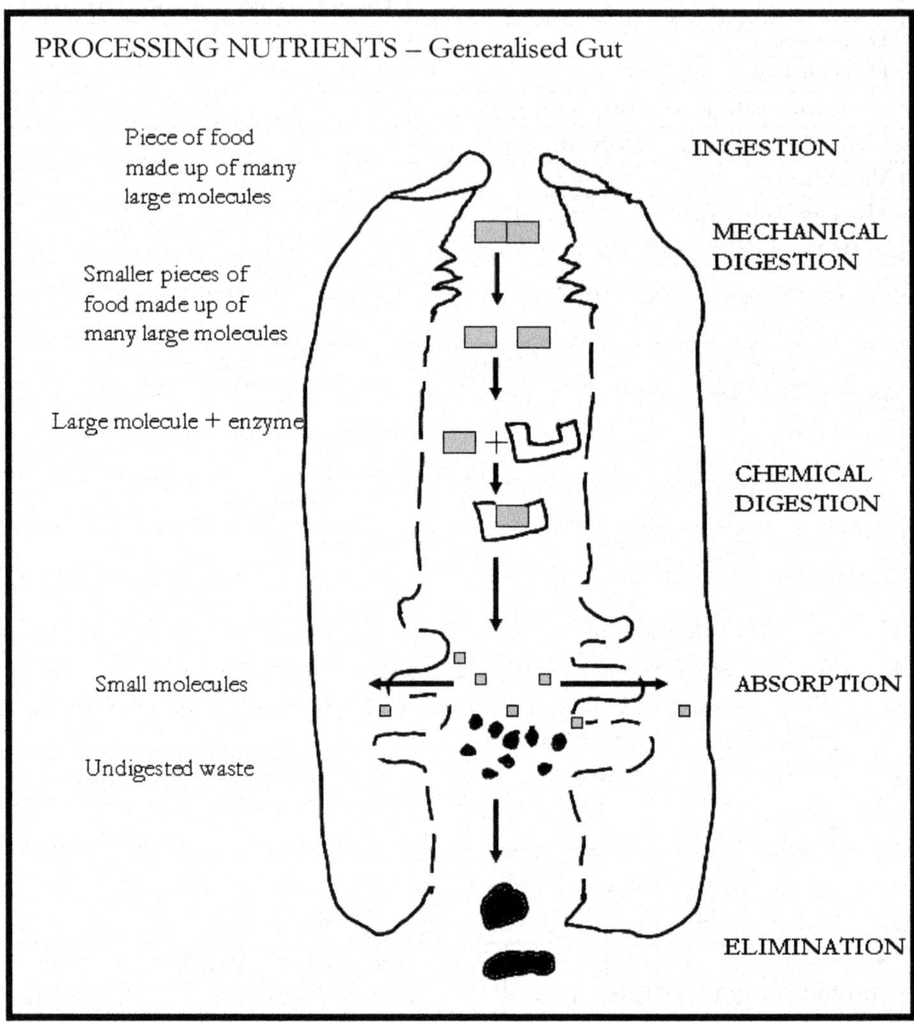

42 Explain how the small intestine is narrow in diameter, yet has a high surface area.

43 Complete the following table.

Food group	Enzyme involved	Product of digestion	Absorbed across the wall of the small intestine into…
Carbohydrate			
Protein			
Lipids			

44 Koalas have a large caecum. Why?

45 Compare the digestive tracts of an herbivore, a carnivore and nectar eater. Consider: teeth, length of digestive tract, stomach and caecum.

46 Construct a table that summarises the source and structures involved in obtaining CO_2, O_2, H_2O, mineral ions and organic compounds in vascular plants.

47 Construct a table that summarises the source and structures involved in obtaining CO_2, O_2, H_2O, mineral ions and organic compounds in a mammal.

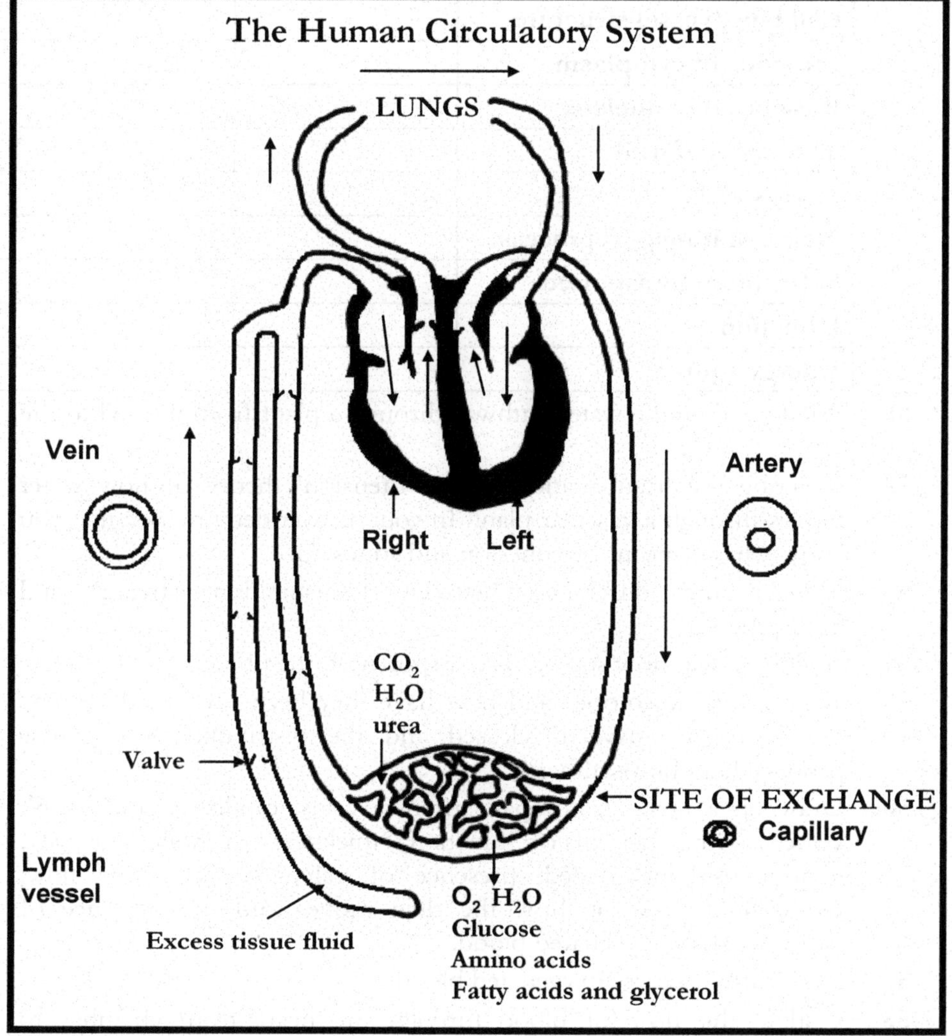

Transport

48 Why do large multicellular organisms require a transport system?

49 What characteristics do effective transport systems share?

50 Transport tissue in vascular plants consists of xylem vessels and phloem cells. Use the table below to summarise transport in vascular plants.

	Xylem	Phloem
Diagram		
Cell type/vessel structure		
Presence of cytoplasm		
Presence of a nucleus		
Thickness of wall		
Alive/dead		
Name of transport process		
Substances transported		
Direction		
Energy source		

51 Briefly outline the water pathway through a plant from the soil to the leaf.

52 Describe the 'transpiration-cohesion-tension' theory of how water moves through a vascular plant. In your answer demonstrate that you know what is meant by cohesion and adhesion.

53 Explain the difference between 'transpiration stream' and 'transpiration'.

54 Briefly explain how sucrose is transported in the phloem.

55 What are radioisotopes and how have they been used to determine the source of oxygen released and the distribution of sucrose produced in photosynthesis?

56 Draw up a table to compare arteries, veins, capillaries and lymph vessels. Consider: major function, thickness of wall, elasticity, continuous/blind ended, presence of valves, direction of flow, blood/fluid pressure, how the fluid moves and the presence of oxygenated/deoxygenated blood.

57 Do all arteries carry oxygenated blood?

58 Outline the flow of blood through the heart from chamber to chamber noting form where the blood comes and its destination.

59 What are the major components of human blood and their function?

60 How are the products of digestion, oxygen, carbon dioxide, nitrogenous waste, ions and hormones transported around the body?

61 Why is it important that oxygen binds reversibly with respiratory pigments?

62 What factors affect how much oxygen can be carried by haemoglobin?

63 Complete the table below summarising the changes in composition of the blood in a capillary bed.

Concentration of	Capillary bed	
	Blood entering	Blood leaving
Oxygen		
Carbon dioxide		
Glucose		
Urea		

64 Insects have an open circulatory system whereas mammals have a closed circulatory system. Explain how these circulatory systems differ. Which is the most efficient?

65 Fish have a single circulatory system whereas mammals have a double circulatory system. Explain how these circulatory systems differ. Which is the most efficient?

66 Compare the roles of the respiratory and circulatory systems by completing the following table.

		Plants	Mammals
Transport Systems	Role		
	Number of separate systems		
	Structures		
	Substances transported		
	Energy source		
	Direction		

Section I Questions

67 Which of the following is a list from most complex to simplest?

(A) colonial, multicellular, unicellular
(B) colonial, unicellular, multicellular
(C) unicellular, colonial, multicellular
(D) multicellular, colonial, unicellular

68 Specialised cells

(A) perform a specific function and always look different from other cells.
(B) perform a specific function and may look different from other cells.
(C) perform a specific function and look the same as other cells in a multicellular organism.
(D) look different from other cells in a multicellular organism but perform the same function as these other cells.

69 The diagrams below are two cells from different multicellular organisms.

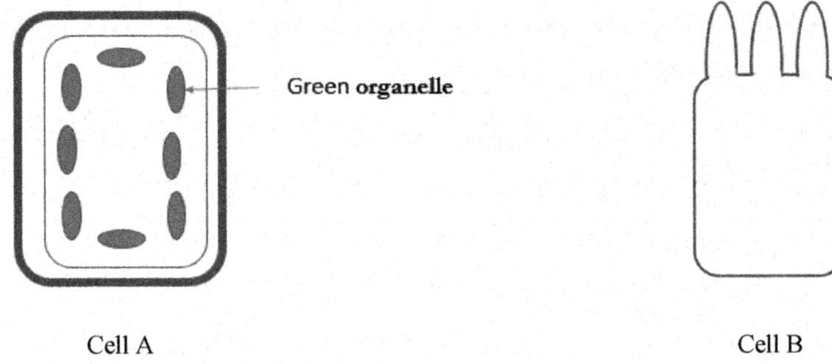

Cell A   Cell B

It is true to say

(A) cell A is differentiated but Cell B is not.
(B) cell B is specialised, but Cell A is not.
(C) both cells are specialised but are not differentiated.
(D) both cells are specialised and differentiated.

70 Which of the following is not a characteristic of an efficient surface for gas exchange?

(A) large surface area
(B) moist surface
(C) a high diffusion gradient
(D) thick epidermal cell membranes

71 If plants are overwatered the soil may become waterlogged and the plant may die. A possible reason for this is

(A) too much water enters the roots causing the root cells to explode.
(B) the plant is unable to take up mineral ions from the soil.
(C) the root cells are unable to get enough carbon dioxide for photosynthesis.
(D) the root cells are unable to get enough oxygen for aerobic respiration and the cells die.

72 Adaptations of leaves to increase photosynthesis include

(A) a thick waterproof cuticle.
(B) a large surface area.
(C) stomata in pits.
(D) hairs on the surface.

73 Within leaves

(A) the palisade layer is the major site of photosynthesis.
(B) epidermal cells usually contain chloroplasts.
(C) the spongy mesophyll is not involved in photosynthesis.
(D) the many air spaces between palisade cells allow for rapid diffusion of gases.

74 The diagram below represents a stoma surrounded by two guard cells. Water is moving from the surrounding epidermal cells into the guard cells.

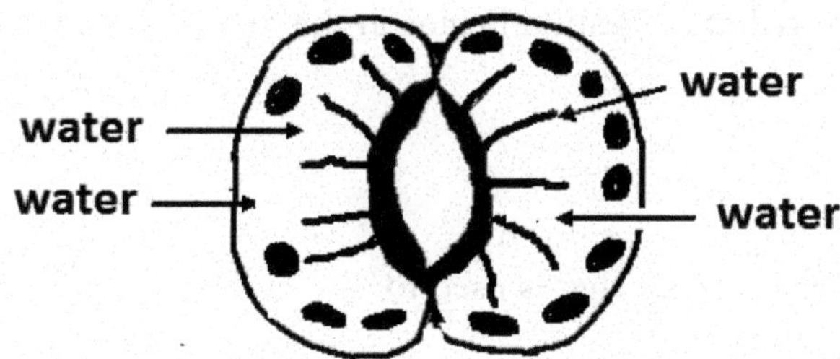

As a result

(A) the size of the stoma will increase.
(B) the size of the stoma will decrease.
(C) less water will be lost from the plant.
(D) photosynthesis will cease in the guard cells.

75 Gas exchange occurs across

(A) alveoli in mammals.
(B) lamellae in insects.
(C) tracheoles in fish.
(D) gills in whales.

76 In the lungs

(A) air is drawn in when the diaphragm relaxes, and the ribs move up and out.
(B) air sacs called alveoli increase the surface area for gas exchange.
(C) oxygen moves into the blood by active transport.
(D) 100% of the oxygen inhaled passes into the blood.

77 For many marine organisms, gas exchange occurs across gills. Gills have

(A) a lower surface area to volume ratio than lung tissue.
(B) a counter current system to increase gas exchange.
(C) a protective barrier to reduce the uptake of salts.
(D) a high surface area to volume ratio to absorb carbon dioxide from the water.

78 A disadvantage of obtaining oxygen from air is

(A) there is less oxygen available in air than in water.
(B) water loss across the respiratory surface may lead to dehydration.
(C) a greater respiratory surface area is required.
(D) oxygen diffuses faster through water.

79 Heterotrophs require digestive systems because

(A) they are unable to build their own organic molecules.
(B) the molecules in the food they eat are too big to pass into the body.
(C) chemical digestion can only occur within cells.
(D) they produce large amounts of waste material.

80 In humans carbohydrate digestion begins in the

(A) mouth.
(B) stomach.
(C) small intestine.
(D) large intestine.

81 Koalas have a large caecum that is used to

(A) store food.
(B) mechanically digest food.
(C) absorb the products of digestion.
(D) digest cellulose.

82 In humans, bile

(A) is involved in the chemical breakdown of lipids.
(B) is involved in the physical breakdown of lipids.
(C) is an enzyme.
(D) is secreted into the stomach.

83 Transpiration will be highest for a well-watered pot plant when the
(A) air is still, and the plant is shaded.
(B) air is moving, and the plant is shaded.
(C) air is still, and the plant is in bright sunlight.
(D) air is moving, and the plant is in bright sunlight

84 With respect to transport within vascular plants, one of the following is true.
(A) It occurs in only one direction.
(B) It occurs only in dead cells.
(C) It requires the sun's energy for movement of water.
(D) Glucose is transported in the phloem.

85 A diagram of a cross-section of a blood vessel is shown below.

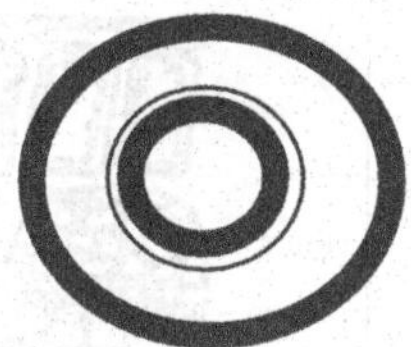

The blood vessel is
(A) a capillary.
(B) a vein.
(C) an artery.
(D) a lymph vessel.

86 Blood returning to the heart via the pulmonary vein
(A) enters the right atrium, flows to the right ventricle and then to the rest of the body via the aorta.
(B) enters the left atrium, flows to the left ventricle and then to the lungs via the pulmonary artery.
(C) enters the left atrium, flows to the left ventricle and then to the rest of the body via the aorta.
(D) enters the left atrium, flows to the right ventricle and then to the rest of the body via the aorta.

87 In the lungs
(A) oxygen diffuses from the blood in the capillary into the alveolus.
(B) oxygen diffuses from the alveolus into the blood in the capillary.
(C) carbon dioxide diffuses from the alveolus into the blood in the capillary.
(D) carbon dioxide diffuses from the capillary into the blood in the alveolus.

88 pH decreases when CO_2 dissolves in water therefore

(A) blood entering a capillary bed will have a lower pH than blood leaving a capillary bed.

(B) blood entering a capillary bed will have a higher pH than blood leaving a capillary bed.

(C) blood entering a capillary bed will have the same pH as blood leaving a capillary bed.

(D) all veins contain blood with a low pH.

Section II Questions

89 Consider the diagrams below.

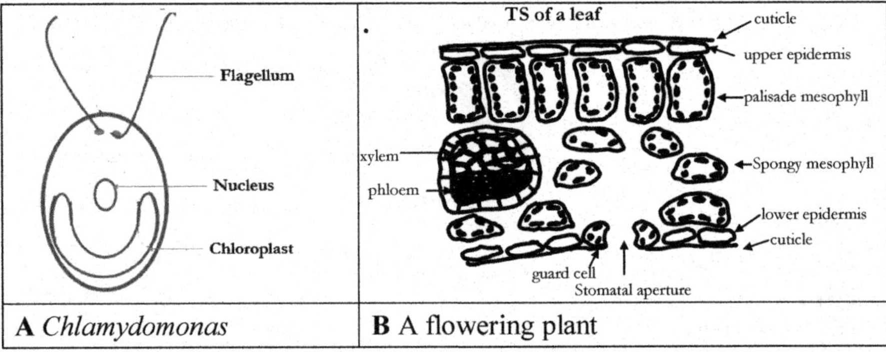

A *Chlamydomonas*	B A flowering plant

(a) Name an organelle apart from the chloroplast that is common to both organisms. [1 mark]

(b) What substances/items are required for photosynthesis by both *Chlamydomonas* and a palisade cell and how do these requirements reach the chloroplast? [3 marks]

(c) Is *Chlamydomonas* more specialised than a palisade cell? Justify your answer. [2 marks]

(d) The leaf is an organ. Name a tissue within the leaf. [1 mark]

[Total 7 marks]

90

(a) In terms of the mammalian lung's function, what are the five major features of the lungs that allow efficient transfer of oxygen? [5 marks]

(b) When exercising, breathing can change in two ways. What are these two ways? Explain how these two changes assist. [3 marks]

(c) What is the function of haemoglobin in red blood cells? [1 mark]

(d) Carbon monoxide is a colourless, odourless gas that can cause unconsciousness and death. Like oxygen it can combine with haemoglobin. Explain how carbon monoxide causes a problem whereas carbon dioxide does not. [2 marks]

[11 marks]

91

(a) What is the role of the human digestive system? [2 marks]

(b) The action of the teeth in the mouth and small intestine differ in terms of the digestive processes. How is digestion different in these two parts of the digestive system? [2 marks]

(c) The small intestine is small in diameter but much longer than the large intestine. Apart from digestion what is the other major role of the small intestine and how is it adapted to carry out this job efficiently? [3 marks]

(d) What is the major role of the large intestine in humans and how is it suited to this function? [2 marks]

[Total 9 marks]

92

During a practical class a student dissected two mammals and compared their digestive systems. The table below summarises the findings.

	Animal R	**Animal S**
caecum	small	large
intestine	long	short

(a) Which animal is an herbivore? [1 mark]

(b) Explain your answer to part **(a)**. [3 marks]

[Total 4 marks]

93

(a) Research and then discuss the use of radioisotopes in the tracing the phloem. [7 marks]

(b) Describe how you gathered and analysed the information in part **(a)**. [3 marks]

(c) How did you evaluate the relevance and reliability of the information gathered in part **(a)**? [2 marks]

[Total 12 marks]

94

(a) What are two major structural differences between arteries and veins that are related to their function? [2 marks]

(b) Capillaries are quite different in structure compared to arteries and veins. How and why? [2 marks]

(c) Explain how the role performed by the pulmonary artery and pulmonary vein are different to that of other arteries and veins? [2 marks]

(d) During foetal development within the uterus, there is a hole between the left side of the heart and the right side of the heart. This hole usually closes at birth. Sometimes this is not the case and sufferers are said to have a hole in the heart. What is the problem and what will be the implications? [2 marks]

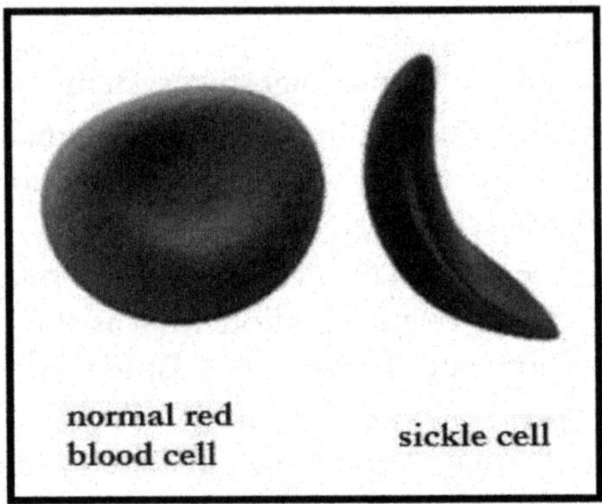

(e) In some Mediterranean countries and Northern Africa, an inherited disease called sickle-cell anaemia is common. Affected individuals have oddly shaped red blood cells. What is the advantage of the shape of normal red blood cells and what are two possible consequences for affected individuals? [3 marks]

[Total 11 marks]

Chapter 4
Biological Diversity
Keywords and Terms:

Effects of the Environment on Organisms

abiotic	abundance	biological control
biotic	distribution	environment
population	selection pressure	

Adaptations

adaptation	behavioural	fauna
flora	homeothermic	physiological
poikilothermic	structural	

Theory of Evolution by Natural Selection

adaptive radiation	allele	allele frequency
analogous	anoxic	biodiversity
convergent evolution	divergent evolution	endosymbiotic
extant	extinct	fitness
gene	gene pool	genotype
geographical barrier	homologous	macroevolution
microevolution	natural selection	organic
oxic	ozone	phenotype
punctuated equilibrium	species	transitional series

Evolution - the Evidence

anatomy	biochemical	biogeography
common ancestor	continental drift	DNA hybridisation
embryology	fossil	Gondwana
homologous	Laurasia	megafauna
Pangaea	phylogenetic tree	radio-isotopic dating
relative dating		

Summary and Content Review Questions

Effects of the Environment on Organisms

Environment

Biotic Factors
Living organisms:
- types (same/different species)
- numbers
- distribution
- interactions (predators/parasites/competition)

Abiotic Factors
Non-living factors:
- temperature
- water/rainfall/humidity
- type of soil/substrate
- salinity
- light
- topography/slope
- wind/wave action
- O_2/CO_2
- mineral ions
- pH

1 Define the terms 'abiotic factors' and 'biotic factors', giving some examples of each.

2 Complete the following table summarising the abiotic factors present in marine, freshwater and terrestrial environments.

Abiotic factor	Environment		
	Marine	Freshwater	Terrestrial
Availability of water			
Availability of light			
Availability gases (oxygen/ carbon dioxide)			
Availability of ions			
Temperature			
Viscosity			
Buoyancy			
Pressure			
Turbulence			
Substrate			

3 Which abiotic factors are the most important in determining the distribution and abundance of aquatic organisms?

4 Which abiotic factors are the most important in determining the distribution and abundance of terrestrial organisms?

5 What biotic factors determine the distribution and abundance of organisms?

6 For a native forest, draw up a list of the biotic and abiotic factors that might affect organisms that live in the area.

7 How does light affect the distribution of photosynthetic organisms (e.g. algae and phytoplankton) in the ocean?

8 Which individuals in a population are most likely to survive in a particular environment?

9 Abiotic and biotic factors are sometimes referred to as selection pressures. What does this mean?

10 The introduction of cane toads into Australia is an example of how a population may change due to selection pressures. Complete the following table.

Scientific name	
General characteristics	
Diet	
Where did the cane toad come from?	
When were they introduced?	
Why were they introduced and was this successful?	
What features of the cane toad allowed the population to increase rapidly?	
Environmental impact	

11 The introduction of prickly pear into Australia is an example of how a population may change due to selection pressures. Complete the following table.

Scientific name	
General characteristics	
Where did the prickly pear come from?	
When were they introduced?	
Why were they introduced?	
What features of the prickly pear allowed the population to increase widely?	
When and how was the prickly pear population controlled?	

Adaptations

Examples of the 3 Adaptation Types

Adaptations are body parts, physiology or behaviour that help an organism to survive in its normal environment

Structural	Physiological	Behavioural
Fins of a fish or hard coral	Counter-current blood flow in dolphin flipper to reduce heat loss	Aggression in male kangaroos

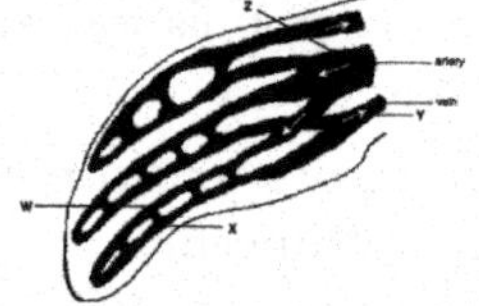

12 Distinguish between structural, physiological and behavioural adaptations.

13 What behaviours can Australian animals use to survive harsh desert conditions?

14 What physiological adaptations can Australian animals use to survive harsh desert conditions?

15 Losing leaves is an adaptation found in some cold climate plants and some desert plants. Explain the significance of the strategy in both sorts of environment.

16 What are the strategies mammals use to gain heat or increase heat loss? Draw up your answers as a table.

17 What behaviours can poikilothermic animals exhibit to regulate body temperature?

18 Homeothermy is thought to have evolved on land even though mammals, such as whales, are homeothermic. What is the advantage of homeothermy to terrestrial mammals and birds?

19 Sweating is a useful way by which many mammals can lose heat. Most desert mammals do not use sweat to cool their bodies. Explain why this is a good strategy and suggest other strategies used to reduce heat gain.

20 What problems do flowering plants have to cope with in hot climates?

21 What adaptations do flowering plants have to cope with hot dry climates?

22 Why is it difficult to know if a feature shown by an organism is an adaptation for living in a particular habitat?

Summary of Water Balance Adaptations

Plants: gaining water	Plants: water loss reduction
• deep growing roots • wider root network • succulent structures to store water when available • plant shape to funnel water to root network	• thick cuticle • leaf colour – lighter & reflective • leaf loss • leaf shape with low surface area to volume ratio • reduced stomata number • stomata in pits • hairs covering leaves • orientation of leaves away from sun
Animals: gaining water	Animals: water loss reduction
• longer large intestine for water absorption from food • greater use of metabolic water • storage of water (when available) as fat • methods of consuming water via plant material and water condensation on leaves early in the morning	• nocturnal – avoiding heat of day • living in humid burrow • body shape with lower surface are to volume ratio • reduced number of sweat glands • kidneys able to produce very concentrated urine • dry faeces

Heat Balance in Animals

Animals: gaining or retaining heat	Animals losing heat
• heat seeking behaviours – e.g. lying in sun • metabolic heat – shivering • reducing blood flow to the extremities • presence of body fat – long term strategy • use of insulating hair/feathers	• Large surface areas – e.g. big ears or body surface • Evaporative cooling - sweating/panting • increasing blood flow to extremities • allowing increase in body temperature during the day (decreases temperature difference between the body and hot environment)

Theory of Evolution by Natural Selection

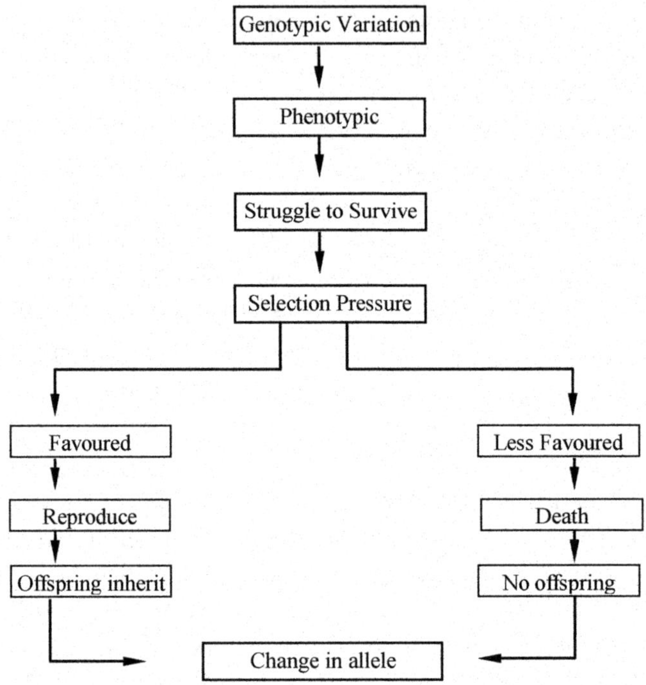

23 What is biological evolution? Is evolution a process or a result? Explain.

24 Is biological evolution a theory or fact?

25 How are 'Natural Selection' and 'Biological Evolution' related?

26 If natural selection is to occur, what must exist in a population?

27 Is selection at the phenotypic or genotypic level? Explain what sorts of individuals are most likely to survive in a particular environment.

28 Does the 'fitness' of an individual vary and if so what determines an individual's fitness?

29 Which individuals in a population are most likely to reproduce in a particular environment?

30 If an individual is to pass on its genetic information to the next generation, how long must it live?

31 Darwin proposed the theory of 'Evolution by Natural Selection'. Since then developments in genetics have led to its modification. Summarise the modern view of evolution by natural selection.

32 Evolution is said to have occurred when there is a change in 'allele frequency' in a 'gene pool'. What is meant by 'allele frequency' and 'gene pool'?

33 Using the Darwin/Wallace theory of natural selection, explain how peppered moths in polluted forests are mainly black whereas populations of these moths in unpolluted forests are mainly pale. This is an example of how physical conditions can affect populations.

34 What is meant by the term ‘species’?

35 What is the most important test used to distinguish members of a species?

36 What is the significance of geographical barriers in the evolution of new species? Describe features that act as geographical barriers.

37 Summarise the important steps that lead to allopatric speciation.

38 How can you determine if you have one species or two distinct species in an area?

39 If two closely related species live in the same area, what are the possible isolating mechanisms that might prevent interbreeding?

40 Is there any real difference in the process of ‘Natural Selection’ when compared with ‘Selective Breeding’?

41 How does selective breeding affect genetic diversity and what are the advantages and disadvantages of this?

42 What is meant by biodiversity?

Approximate time and sequence of life on Earth (*Times in millions of years)

Time*	Environment	Life	Evidence
Now			
500		land colonisation	
1000			
1500		multicellular eukaryotes	fossils/molecular clocks
		unicellular eukaryotes	fossils
2000	ozone accumulates	aerobic prokaryotes	fossils/red bed formations
2500	increase in oxygen		banded iron formations
3000			
		photosynthetic prokaryotes	fossils (stromatolites)
3500	water cycle	Anaerobic, heterotrophic prokaryotes	fossils sedimentary rock dating
4000			
	Earth cools		
4600	Earth formed		mineral crystals dating

43 Approximately how old is the Earth?

44 Describe conditions on early Earth.

45 How and when was the water cycle established? On what evidence is this based?

46 What is meant by 'organic molecule'?

47 What elements were necessary before life could form?

48 What two other factors were essential for life on Earth?

49 Describe the Urey and Miller experiments. What were their findings and what is the significance of these findings?

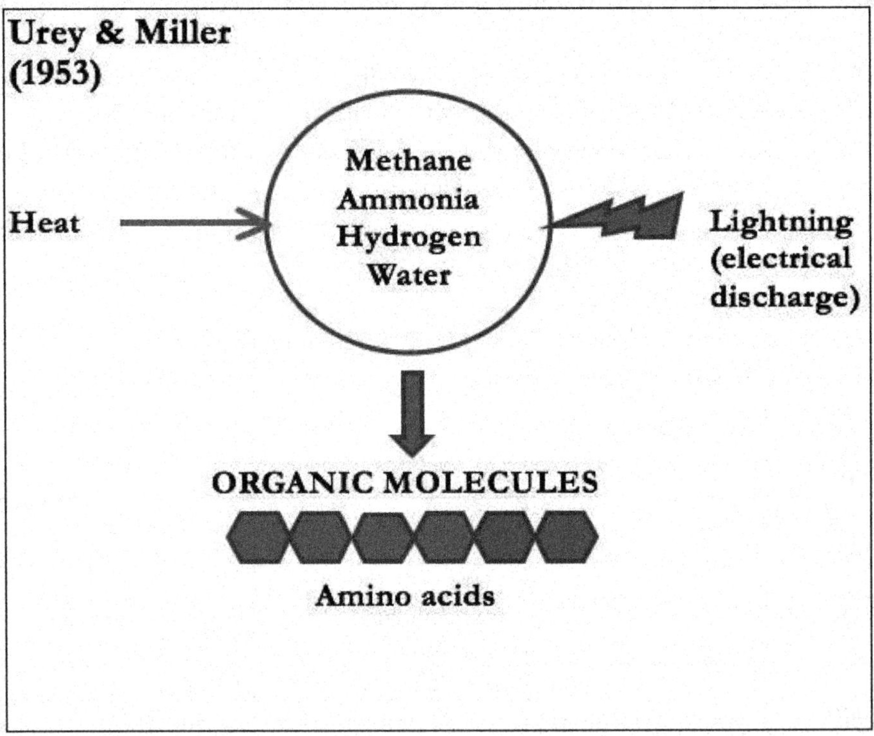

50 The atmosphere of early Earth lacked oxygen. How would this have helped the formation of the first organic molecules?

51 What steps would need to happen before first organic molecules could become cells?

52 Approximately when did life develop on Earth?

C O H N P S

↓

Small organic molecules

e.g. amino acids nucleotides

↓

Large organic molecules

e.g. Proteins, DNA

↓

Single cells

Heterotrophic prokaryotes

Autotrophic prokaryotes

Aerobic prokaryotes

Eukaryotes

↓

Colonies of cells

↓

Multicellular organisms

53 Complete the following summary of the major stages in the evolution of living things.

Major stage in the evolution of life	Description
1 formation of organic molecules	
2 formation of membranes	
3 formation of prokaryotic heterotrophic cells	
4 formation of prokaryotic autotrophic cells	
5 formation of aerobic prokaryotic cells	
6 formation of eukaryotic cells	
7 formation of colonial organisms	
8 formation of multicellular organism	

54 Describe the palaeontological evidence that suggests when life originated on Earth.

55 Describe the geological evidence that suggests when life originated on Earth.

56 When and how was oxygen added to Earth's atmosphere? What evidence supports this?

57 What is the endosymbiotic theory of evolution of eukaryotic cells and what is the supporting evidence?

58 What is the advantage of aerobic respiration over anaerobic respiration to cells?

59 How was the ozone layer formed and why was this significant for life on Earth?

60 Draw a timeline using the information in the following table.

	Approximate time
First aerobic microbes	2500 m.y.a.
Earliest reptiles	330 m.y.a.
First flowering plants	130 m.y.a.
Earth formed	4500 m.y.a.
First multicellular eukaryotes	600 m.y.a.
First amphibians	365 m.y.a.
First insects	479 m.y.a.
First conifers	250 m.y.a.
First prokaryotic cells	3800 m.y.a.
First vertebrates	500 m.y.a.
First mammals	160 m.y.a
First ferns	360 m.y.a.
First photosynthetic cells	3500 m.y.a.
First land plants	450 m.y.a.
First unicellular eukaryotic	2100 m.y.a.

61 The fossil record of the evolution of horses forms a transitional series. What does this mean?

62 The ancestor of horses arose around 52 million years ago. What did this ancestor look like?

63 Charles Darwin visited Australia (NSW, Tasmania and Western Australia) in 1836. What were his two major observations about the flora and fauna of Australia?

64 In 1836, Charles Darwin observed a platypus in the Blue Mountains. It was clear that the platypus was a mammal, but it shared many characteristics with reptiles. Summarise these observations by completing the following table.

Reptile-like	**Mammal-like**	**Unique to the platypus**

65 Use the finches of the Galapagos Islands to explain what is meant by adaptive radiation.

66 Name two features that characterise adaptive radiation.

Comparison of Divergent and Convergent Evolution

	Divergent (Adaptive Radiation)	Convergent
	Organism 1 ↖ ↗ Organism 2 Common ancestor	Organism 1 ↑ Ancestor 1; Organism 2 ↑ Ancestor 2
Appearance	Different	Similar
Recent common ancestor	The same	Different
Environment	Different	Same
Evidence	Homologous structures	Analogous structures
E.g.	Australian marsupials	Flying squirrels and sugar gliders

67 Distinguish between divergent and convergent evolution. Include an example of each.

68 What is another name for divergent evolution?

69 Is convergent or divergent evolution more likely to lead to evolution of new species? Explain.

70 Can individual organisms become extinct? Explain.

71 Describe what is meant by the 'punctuated equilibrium' model of evolution.

72 How does 'punctuated equilibrium' differ from natural selection?

Evolution – the Evidence

73 Complete the following table.

	Evidence for evolution	Example
Biochemical (molecular homology)		
Comparative anatomy		
Comparative embryology		
Biogeography		
Palaeontology		

74 What is molecular homology and how can it be used to determine the relatedness between species?

75 How is mitochondrial DNA different from nuclear DNA?

76 Why is mitochondrial DNA useful for tacking human evolution?

77 What is DNA hybridisation and how can it be used to determine relatedness between species?

78 Draw a phylogenetic tree based on the following hybridisation results.

Human and chimpanzee: 97.7%

Human and gibbon: 94.7%

79 How do analogous and homologous structures differ?

Pangaea

Laurasia	**Gondwana**
North America	South America
Europe	Africa
Asia	Madagascar
	India
	Antarctica
	Australia
	New Zealand
	New Guinea

80 In 1912, Alfred Wegener proposed the Theory of Continental Drift. What is meant by 'Continental Drift'?

81 In the 1960s Harry Hess developed the Theory of Plate Tectonics. Explain how this theory provides a mechanism for continental drift.

82 Name the continents that exist on Earth today.

83 What was the name of the super-continent formed by Earth's landmasses about 260 million years ago?

84 When did Pangaea split into Laurasia and Gondwana?

85 What present day continents were once parts of Laurasia?

86 Draw a timeline showing the breakup of Gondwana.

87 At what rate is Australia moving north?

88 Complete the following summary of the evidence suggesting Australia was once part of Gondwana.

Evidence	Description
Matching continental margins	
Position of mid-ocean ridges	
Spreading zones between continental plates	
Fossils in common on Gondwanan continents	
Similarities between present day organisms on Gondwanan continents	

89 What is meant by the term 'megafauna?'

90 When did the Australian megafauna become extinct? What are two possible explanations of the extinction?

91 What is meant by 'extant species'?

92 The lung fish is a living fossil. What does this mean? Give further examples.

93 Is it always true to say that fossils are the remains of dead organisms?

94 Under what environmental conditions are fossils formed? What conditions reduce the chances of good fossils forming?

95 What information can fossils provide about the lifestyle of once living organisms?

96 Why is it difficult to identify different species using only fossil evidence?

97 Describe the difference between absolute dating (radioactive isotope dating) and relative dating (stratigraphy).

98 Different radioactive isotopes have different half-lives. What does this mean?

99 Different radio-isotopic methods are used to date fossils of different ages. List some commonly used isotopes and state their use. Is it always the fossil that is dated?

100 What is the significance of fossils like *Archaeopteryx*?

101 Cane toads continue to spread throughout Australia. How do the toads leading the spread or expansive front differ from their ancestors?

102 Explain why cane toads not on the expansive front are more sedentary than those on the expansive front.

103 Using the Darwin/Wallace theory of natural selection, explain how mosquitoes have become resistant to DDT. This is an example of how chemical conditions can affect populations.

Section I Questions

104 All of the following are abiotic factors except

(A) air temperature.
(B) soil particle size.
(C) humus content of soil.
(D) water holding capacity.

105 All of the following are biotic factors except

(A) dead remains of organisms.
(B) the concentration phosphates in the soil.
(C) density of rabbit burrows in an area.
(D) distribution of bird nests.

106 Abiotic factors that affect living organisms in the marine environment are

(A) concentration of phytoplankton per litre of water.
(B) salinity of the water.
(C) concentration of ammonia.
(D) broadcast spawning.

107 Living on land or in water presents different problems to the organisms that exist in different environments. In comparing problems of living on land to living in water, which of the following is correct?

(A) More oxygen is available to organisms that live in water.
(B) Water offers more support than air to aquatic animals.
(C) Light penetrates the water, so plants can survive in deep water.
(D) There is a greater risk of desiccation in aquatic environments.

108 The cane toad was introduced into Australia as a biological control for cane beetles. Control of cane beetles was unsuccessful because

(A) not enough cane toads were introduced.
(B) cane toads were unable to breed.
(C) cane toads cannot jump very high, so the cane beetles were able to retreat to the top of the sugar cane.
(D) cane toads could not digest cane beetles.

109 Cane toads are not a problem in Hawaii and South America because

(A) climatic conditions are not suitable for their reproduction.
(B) not enough food is present.
(C) more water is available.
(D) predators are present.

110 In Australia the prickly pear population has been brought under control by the introduction of *Cactoblastis cactorum.* This is an example of how a population may change due to a selection pressure. The selection pressure in this example is

(A) the presence of *Cactoblastis cactorum.*
(B) the presence of prickly pear.
(C) the presence of man.
(D) the climate.

111 Structural adaptations include the following

(A) a cheetah's ability to run extremely fast
(B) an elephant using its trunk to cool itself
(C) tough waterproof eggshell to protect bird embryo
(D) stomach wall that secretes acid to aid digestion

112 Behavioural adaptations include the following:

(A) the cheetah's ability to run extremely fast
(B) the camel's ability to survive long periods without water
(C) tough waterproof eggshell to protect bird embryo
(D) the desert hopping mouse's nocturnal habits

113 Shelled eggs are an adaptation to living on land because they

(A) allow gas exchange between the embryo and the environment.
(B) they protect the embryo from mechanical damage.
(C) they reduce exposure of the embryo to light.
(D) they prevent dehydration of the embryo.

114 Plant leaves of different species have features that allow them to survive in their particular environment. It is reasonable to suggest that

(A) water lilies have more stomata on the bottom of their leaves as water loss is a negligible.
(B) the hanging leaves of gum trees reduces water loss.
(C) leaves with stomata in pits are usually found in rainforests.
(D) plants with ephemeral life cycles are suited to areas with good rainfall.

115 A plant adaptation commonly found in dry climates is

(A) a thin waxy cuticle.
(B) small numerous stomata.
(C) plants with an inverted stomatal rhythm.
(D) ability to lose leaves during wettest part of the year.

116 Desert animals, such as the desert hopping mouse, have a variety of adaptations to overcome a lack of regular water supply such as

(A) sheltering in a humid burrow at night.
(B) production of large amounts of concentrated urine.
(C) increased numbers of sweat glands to increase heat loss.
(D) greater reliance on the use of metabolic water.

117 An animal's body can gain or lose heat depending on the temperature difference between the animal's body temperature and the temperature of the environment. One of the following is correct.
(A) Heat gain via evaporation is small in cold conditions.
(B) Animals can gain heat by conduction in cold conditions.
(C) Animals can gain or lose heat by convection.
(D) Heat loss by evaporation increases in humid conditions.

118 In cold environments animals can increase heat production by
(A) restricting blood flow to the body's extremities (e.g. tips of ears).
(B) raising body hairs to trap heat.
(C) drinking warm fluids.
(D) increasing their rate of shivering.

119 In hot environments animals can increase heat loss by
(A) increasing blood flow to the body's extremities (e.g. tips of ears).
(B) raising their metabolic rate.
(C) reducing sweating to conserve water.
(D) reducing thickness of fat layer.

120 Which of the following populations would be least likely to survive a change in environment?
(A) a population of asexually reproducing bacteria
(B) a population that has been bred from individuals from small endangered populations
(C) an island population of individuals descended from a small migrant population
(D) a population of individuals showing many different phenotypic variations

121 Biodiversity refers to
(A) the variety plants on Earth.
(B) the variety of animals on Earth.
(C) the variety of prokaryotes on Earth.
(D) the variety of all life on Earth.

122 Biodiversity has been decreased by
(A) clearing of forests.
(B) global warming.
(C) agriculture.
(D) all of the above.

123 The age of the Earth has been estimated by scientists as approximately
(A) 10 000 years old.
(B) 6 500 000 years old.
(C) 4500 million years old.
(D) 300 million years old.

124 The early atmosphere of the Earth contained
(A) high levels of free oxygen.
(B) high levels of carbon dioxide.
(C) high levels of ozone.
(D) high levels of ammonia.

125 The scientists Alexander Oparin and John Haldane were the first to
(A) suggest the chemicals for life came from outer space.
(B) suggest the chemicals for life came from Earth.
(C) demonstrate that inorganic molecules could be converted into organic molecules under conditions that existed on early Earth.
(D) demonstrate that organic molecules could be converted into inorganic molecules under conditions that existed on early Earth.

126 Organic molecules contain
(A) carbon, hydrogen and ozone.
(B) carbon, hydrogen and oxygen.
(C) carbon, nitrogen and iron.
(D) hydrogen and oxygen.

127 The experiments of Harold Urey and Stanley Miller
(A) proved that life on Earth came from outer space.
(B) showed that life on Earth could have come from outer space.
(C) proved that life on Earth originated on Earth.
(D) showed that life on Earth could have originated on Earth.

128 Harold Urey and Stanley Miller recreated the atmosphere of early Earth in the laboratory and produced
(A) amino acids.
(B) proteins.
(C) nucleic acids.
(D) fatty acids.

129 Conditions required for the formation of organic molecules in the laboratory include
(A) oxygen.
(B) low temperatures.
(C) electric discharge.
(D) light.

130 Which of the following shows the generally accepted order beginning with the oldest, of the major stages in the evolution of living things?
(A) aerobic prokaryotes, anaerobic prokaryotes, photosynthetic prokaryotes, eukaryotes
(B) anaerobic prokaryotes, photosynthetic prokaryotes, aerobic prokaryotes, eukaryotes
(C) eukaryotes, photosynthetic prokaryotes, aerobic prokaryotes, anaerobic prokaryotes
(D) photosynthetic prokaryotes, eukaryotes, anaerobic prokaryotes, aerobic prokaryotes

131 What was the energy source for the first forms of life?
(A) the sun
(B) ingestion of simple organic molecules from their surroundings
(C) ozone
(D) other eukaryotic cells

132 The endosymbiotic theory suggests
(A) mitochondria were once free-living prokaryotes.
(B) prokaryotes were formed when cells engulfed eukaryote cells.
(C) ribosomes were once free-living prokaryotes.
(D) multicellular organisms resulted from a mutually benefitted association of single cells.

133 Below is a list of stages that needed to occur before organic molecules could form cells.

- **i** the formation of self-replicating molecules egg DNA
- **ii** the formation of small organic molecules
- **iii** the packaging of organic molecules in a membrane
- **iv** the formation of organic polymers from small organic molecules

The correct order from first to last is
(A) i, ii, iii, iv.
(B) iii, iv, i, ii.
(C) iv, i, iii, ii.
(D) ii, iv, i, iii.

134 The atmosphere of early Earth contained very little free oxygen. Gradually oxygen was added to the atmosphere by
(A) heterotrophic prokaryotes.
(B) heterotrophic eukaryotes.
(C) photosynthetic autotrophs.
(D) chemosynthetic autotrophs.

135 The following is an example of divergent evolution:
(A) North American flying squirrels and Australia's sugar glider
(B) the similar digestive physiologies of deer and kangaroos
(C) Australian marsupials present in rainforests, grasslands and open woodlands
(D) the similar appearance of the placental mole and the marsupial mole

136 The different size and shape of the beaks of different species of finches on the Galapagos Islands is an example of adaptive radiation. The selection pressure involved is
(A) the type of food source.
(B) the type of predator present.
(C) the amount of rain.
(D) the different types of finches.

137 Punctuated equilibrium differs from natural selection in that

(A) punctuated equilibrium is associated with animals whereas natural selection is associated with plants.

(B) punctuated equilibrium is associated with plants whereas natural selection is associated with animals.

(C) punctuated equilibrium involves slow change whereas natural selection involves rapid change.

(D) punctuated equilibrium involves rapid change whereas natural selection involves slow change.

138 The table below shows the number of amino acid differences in a haemoglobin chain between humans and a range of animals.

Animal	**Number of different amino acids in haemoglobin compared with a human**
Gorilla	1
Rhesus monkey	8
Mouse	27
Frog	67

Using this data, which animal is most distantly related to humans?

(A) gorilla

(B) rhesus monkey

(C) mouse

(D) frog

139 The diagram below shows the evolutionary relationship between three animals based on differences in amino acid sequences of cytochrome c.

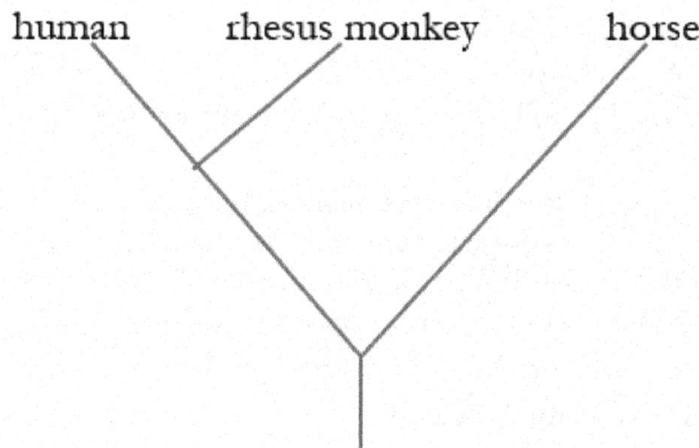

Based on this information, it may be concluded that

(A) humans are more closely related to rhesus monkeys than horses.

(B) rhesus monkeys are more closely related to horses than humans.

(C) humans are descended from rhesus monkeys.

(D) humans and horses share the most recent common ancestor.

140 Amphibians, reptiles, birds and mammals have pentadactyl limbs this suggests

(A) they have four digits on each limb.
(B) they are unrelated in terms of evolution.
(C) they live on land.
(D) they share a common ancestor.

141 The name of the supercontinent that formed about 260 million years ago is

(A) Pangaea.
(B) Laurasia.
(C) Gondwana.
(D) Antarctica.

142 The following were part of Laurasia:

(A) North America, Europe and India.
(B) South America, Africa and Australia.
(C) South America, Africa and Antarctica.
(D) North America, Europe and Asia.

143 The following were part of Gondwana:

(A) South America, Asia and Africa.
(B) South America, Africa and Australia.
(C) North America, Europe and India.
(D) North America, Europe and Asia.

144 The following will most likely occur where continental plates move apart:

(A) uplifting of mountain ranges.
(B) formation of new sea floor.
(C) formation of glaciers.
(D) formation of deserts.

145 Which of the following provides support for the existence of Gondwana?

(A) The presence of the dingo in Australia.
(B) the large number of species endemic to Australia.
(C) the worldwide distribution of trilobite fossils.
(D) the distribution of ratite birds across the southern hemisphere.

146 Gondwana began to break up with a split between

(A) South America and Africa.
(B) New Zealand and Antarctica.
(C) South America and Antarctica.
(D) Australia and Antarctica.

147 The Australian megafauna
(A) included a giant cow.
(B) became extinct before indigenous Australians arrived in Australia.
(C) may have become extinct as a result of climate change.
(D) were well adapted to tropical climates.

148 The flightless ratite birds are only found in the southern hemisphere. It is thought that they shared a common ancestor. Their distribution
(A) is evidence that Gondwana once existed.
(B) similar environments have selected for flightlessness.
(C) is evidence that Laurasia once existed.
(D) is a result of them arising independently in more than one area.

149 Which of the following best describe the vegetation of Gondwana about 150 million years ago?
(A) conifers, cycads, ferns
(B) conifers, cycads, ferns and flowering plants like *Nothofagus*
(C) grasslands and open forests
(D) Eucalypt forests

150 Sclerophyll plants
(A) require large amounts of water and high temperatures.
(B) have small leaves resulting in reduced water loss.
(C) have large leaves resulting in reduced water loss.
(D) have large leaves resulting in increased absorption heat from the environment.

151 Changes in Australian fauna and flora is thought to be due to
(A) climate change.
(B) continental drift.
(C) human intervention.
(D) all of the above.

152 It is thought that life originated on Earth about 3850 million years ago. Supporting palaeontological evidence includes
(A) the presence of banded iron formations in sedimentary rock.
(B) a higher ratio of ^{12}C to ^{13}C in ancient rocks.
(C) fossils dating back 3500 million years.
(D) increasing atmospheric oxygen content over time.

153 The diagrams below represent layers in sedimentary rock from two different locations.

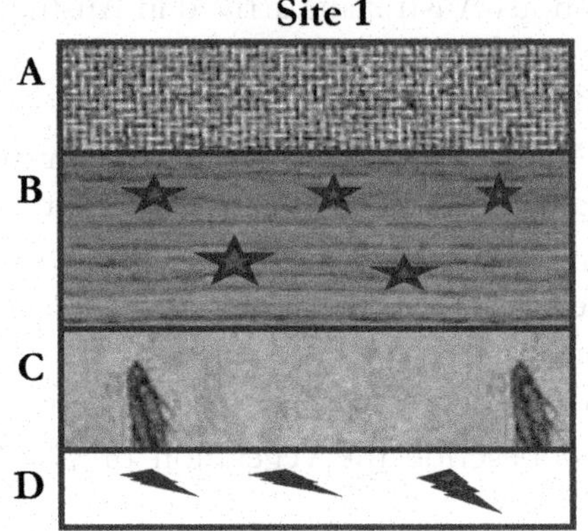

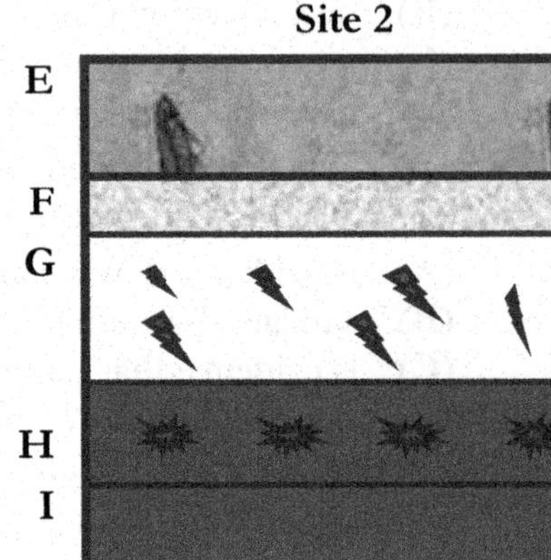

These diagrams suggest that fossils found

(A) in layer I are younger than fossils found in Strata B.
(B) in layers I and D are the same age.
(C) in layer D are older than fossils found in layer H.
(D) in layers C and E are the same age.

154 Insects in amber are examples of

(A) indirect fossils.
(B) direct fossils.
(C) castes.
(D) moulds.

155 ^{14}C has a half-life of approximately 6000 years. What percentage of the original ^{14}C would remain in a 12 000-year-old fossil?

(A) 100%
(B) 50%
(C) 25%
(D) 0%

156 The oldest dingo fossil has been dated at 3450 years old using

(A) ^{14}C dating of the fossil bones.
(B) ^{14}C dating of the surrounding rock.
(C) potassium–argon dating of the fossil bones.
(D) potassium–argon dating of the surrounding rock.

157 Very few fossils of jellyfish have been found because

(A) they have only recently evolved.
(B) they live in water.
(C) they are too small.
(D) they are not made of hard structures.

Section II Questions

158

The rocky shore presents a number of challenges for organisms that inhabit the intertidal zone. Some of the factors are abiotic and others have biotic causes. Different organisms live in different zones or bands on the shore. The zones are often visually obvious as stripes of organisms across the platform. For example, bands of mussels (a bivalve) or barnacles (sessile arthropods) are often seen at different positions on the shore.

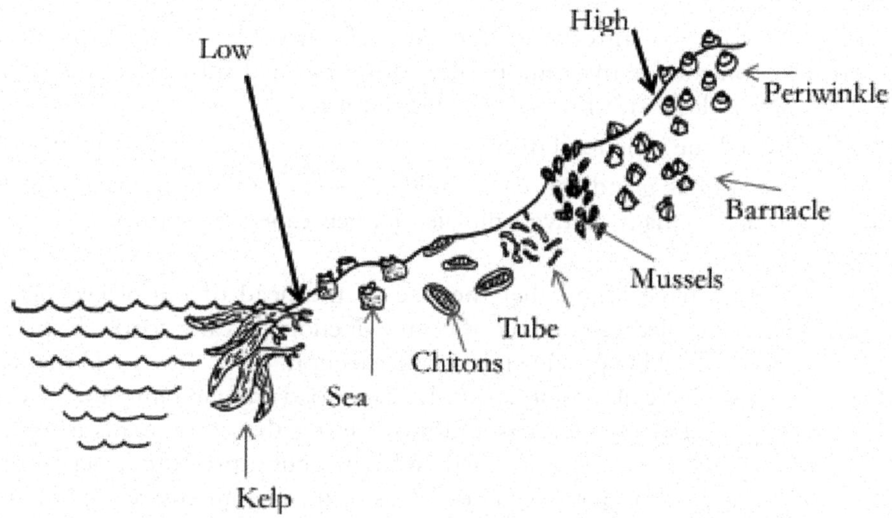

(a) List three abiotic factors that have an impact on organisms on a rocky shore and name the instrument you would use to measure the factor.

[3 marks]

(b) What is the major biotic factor that determines the distribution of organisms on a rock platform? Illustrate your answer with an example of the types of organisms involved. [2 marks]

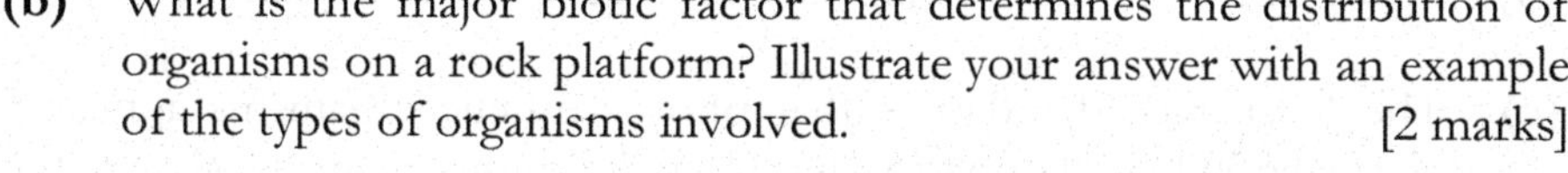

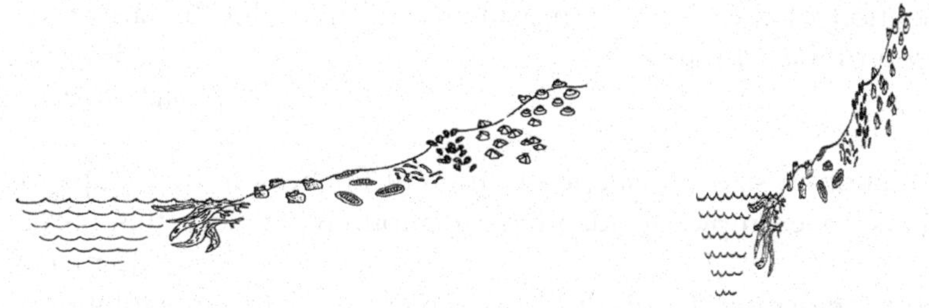

Flat rock platform

Steep rock platform

(c) How can the slope of the shore have an impact on the zonation pattern on a rocky shore? [2 marks]

(d) Conditions vary across the rock platform. Describe two sites with different conditions that are commonly found on a rock platform. [2 mark]

(e) Describe the adaptations of some of the larger animals that are capable of movement, found on the rocky shore. One example should be of behaviour and the other of structural adaptation. [2 mark]

[Total 11 marks]

159

The cane toad was introduced into Australia from Hawaii in 1955 as a biological control for the cane beetle. The cane toad thrived in Australia, but it had very little effect on cane beetle numbers.

(a) Why did the cane toad thrive? [2 marks]

The cane toad has glands on either side of its head that produce a poison that is toxic to many native animals. Larger cane toads produce more poison.

(b) Scientists have found that the size of the head of red-bellied black snakes has decreased since the introduction of cane toads. Explain how this could occur due to natural selection. [3 marks]

(c) Scientists have also found that the cane toads moving into new areas are larger, have longer legs and move faster than cane toads in areas where they are well established. What selection pressure is acting on the cane toads moving into new areas. Explain your answer [2 mark]

[Total 7 marks]

160

The first forms of life on Earth were anaerobic, heterotrophic prokaryotes.

(a) Why do scientists think the first forms of life were anaerobic? [1 mark]

(b) From where did they obtain the energy required to sustain life? [1 mark]

(c) The fossil record indicates that photosynthetic organisms appeared just over 3000 million years ago. How did these organisms change the composition of the Earth's atmosphere and how did this affect the evolution of life on Earth? [3 marks]

[Total 5 marks]

161

(a) The variation in size and shape of different finches on the Galapagos Islands is an example of adaptive radiation. What does this mean? [2 marks]

(b) Is adaptive radiation the result of convergent or divergent evolution? [1 mark]

[Total 3 marks]

162

Modern classification systems reflect evolutionary relationships. These are usually based on molecular comparisons.

(a) Explain how molecular comparisons can be used to show evolutionary relationships. [2 marks]

(b) In the past molecular comparisons of fossils has been difficult. Describe how one new technology has made molecular comparison of fossils easier. [2 marks]

(c) Comparison of the order of amino acids in cytochrome c can be used to determine the relatedness of different species. The table below shows the number of differences in cytochrome c for three different species.

		Species B	Species C
	Species A	22	6
	Species B		20

(d) Using this information, draw a phylogenetic tree to show the evolutionary relationship between the three species. [1 mark]

(e) Explain why structural comparisons are not always a good indication of evolutionary relationships. [3 marks]

[Total 8 marks]

163

Very few platypus fossils have been found and almost all have been found in Australia. The oldest fossil is 110 million years old and is an opalised jaw fragment of *Steroprodon galmani.* This fossil was found at Lightning Ridge in northwest New South Wales. A fossilised complete skull of *Obdurodon dicksoni* was found at Riversleigh in north Queensland. It has been dated at 10–23 million years old. Only one platypus fossil has been found outside Australia. It was found in 1991 in southern Argentina and its age has been estimated at 61–63 million years old.

(a) Many scientists have suggested that Australia was once part of a landmass called Gondwana. Is evidence supporting this, provided by the platypus fossil discoveries? Explain your answer. [3 marks]

(b) Based on the above information where did the platypus evolve and how was it dispersed? Justify your answer. [3 marks]

(c) Give two reasons why platypus fossils have not been found in Africa. [2 marks]

[Total 8 marks]

164

Megafauna fossils have been found at Riversleigh in north Queensland and Naracoorte South Australia.

(a) What is meant by 'megafauna'? Give one example. [2 marks]

(b) Many scientists suggest the Australian megafauna became extinct as a result of human interference. How could human activity result in the extinction of the megafauna? [2 marks]

(c) State one piece of evidence that supports extinction as a result of human interference and one piece of evidence against this theory. [2 marks]

[Total 6 marks]

165

A fossil of an extinct flightless bird called 'progura' has been found in Leaena's Breath cave near the border of WA and NSW in the Nullarbor Plain. The cave has been called an avian graveyard because of the number of bird remains that have been found in it. When high pressure systems move across the cave's entrance air rushes into the cave at high speeds (up to 70 km/h). Any birds in flight are blown into the cave.

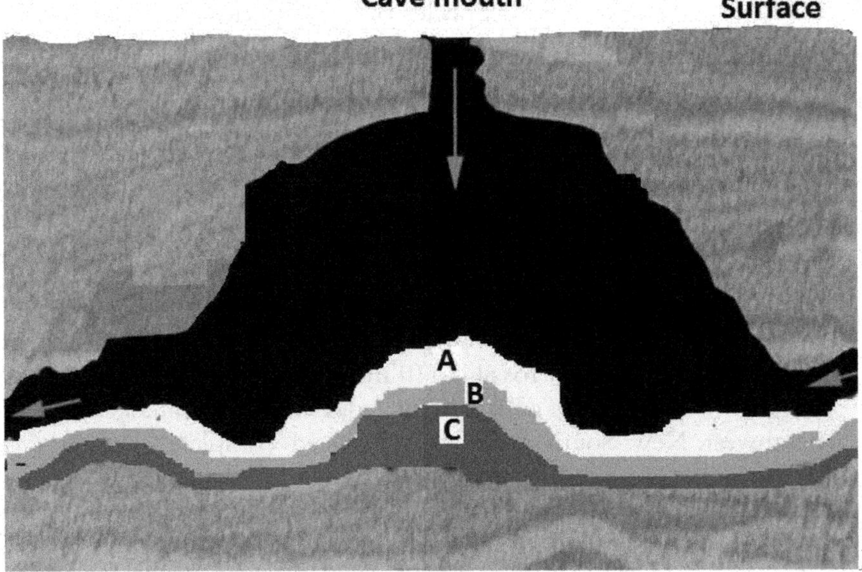

Figu 4.7

(a) Describe two features of the cave that aid the formation and preservation of fossils. [2 marks]

(b) Scientists estimate that the progura fossil is 780 000 years old. Outline how the scientists would have dated the fossil. [2 marks]

(c) At first scientists thought the progura fossil was of a wedge-tailed eagle. What would scientists do to determine that the fossil was not a wedge-tailed eagle? [1 mark]

(d) Fossils have been found in layers A, B, and C. Which layer would contain the oldest fossils? [1 mark]

[Total 6 marks]

166

A nearly complete skeleton of *Diprotodon optatum* has been found on the Leichhardt River between Burketown and Normanton. This *Diprotodon* is estimated to have lived 100 000 to 200 000 years ago.

(a) What method would have been used to date this fossil? [2 marks]

Diprotodons are distant relatives of wombats. The fossil found is of an elderly specimen that would have weighed 3 tonnes and stood 2 metres high.

(b) How would scientists estimate the size and weight of the specimen from its fossilised bones? [2 marks]

(c) Relatively few fossils of *Diprotodons* have been found in Northern Australia. Suggest one reason for this. [1 mark]

(d) Some broken teeth from giant predator lizards were found amongst the fossil *Diprotodon* bones. What does this suggest about how the *Diprotodon* died? [1 mark]

[Total 6 marks]

Chapter 5
Ecosystem Dynamics
Keywords and Terms:

Population Dynamics

allelopathy, amensalism, capture-recapture, carnivore, carrying capacity, commensalism, community, competition, decomposer, disease, ecosystem, ecosystem, exponential, food chain, food web, habitat, herbivore, host, interspecific, intraspecific, microhabitat, mutualism, niche, omnivore, parasitism, predation, predator, prey, quadrat, symbiosis, tolerance range, transect

Past Ecosystems

arboreal, arid, biological control, half-life, ice core, radiometric dating, sclerophyll

Future Ecosystems

bioindicator, habitat fragmentation, habitat loss, revegetation, scientific model

Summary and Content Review Questions

Population dynamics

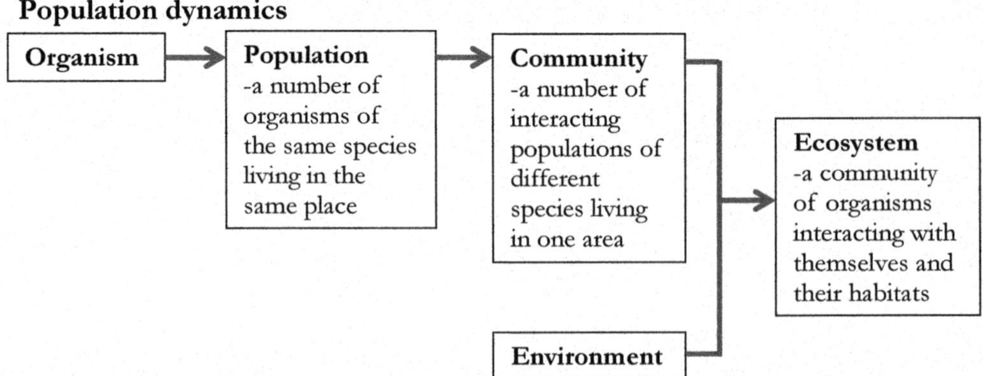

1 Distinguish between habitat, community and ecosystem.

2 Distinguish between the terms 'habitat' and 'microhabitat'.

3 Some species have a very narrow geographic range. Others occupy a broad diverse area. Explain how this is possible.

4 Sketch and label a graph to illustrate the tolerance range concept. Include the optimum range, zone of physiological stress and zone of intolerance.

5 Distinguish between the terms population and community.

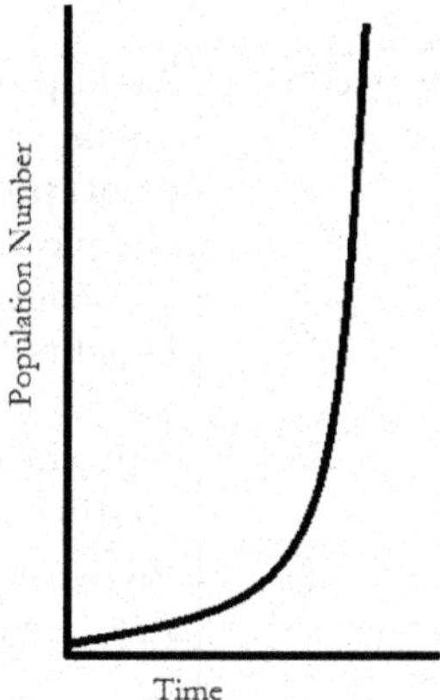

Exponential Growth

6 What is meant by 'exponential growth'?

7 Most natural populations do not exhibit exponential growth. Usually they reach a peak and then oscillate around a maximum sustainable carrying capacity. What is the carrying capacity determined by?

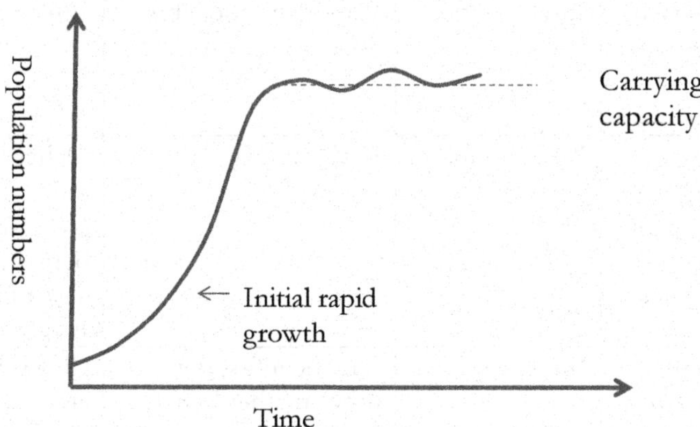

Population Growth Curve

8 What often happens to populations that exhibit uncontrolled exponential growth? Why?

9 Interactions between organisms can be intraspecific or interspecific. What does this mean? Include examples in your answer.

Interactions within a Community			
Interaction	**Intra/ Interspecific**	**Comment**	**Example**
Competition - common demand for limited resources by 2 or more organisms	Intraspecific		Competition between birds of the same species for nesting sites
	Interspecific	**Allelopathy** – an organism (usually plant) produces a chemical that influences (harms or benefits) the growth, survival or reproduction of another organism	No plant growth under pine trees that are covered in pine needles
Predator–Prey - when one species eats another	Interspecific	Predator – benefits Prey – harmed	Kookaburra (predator) eats lizard (prey)
Parasitism - one species lives on or in another species (host), deriving food from it but usually not killing it	Interspecific	Parasite – benefits Host – harmed	Tapeworms living in dogs
Infectious disease - one species impairs the functioning of another species	Interspecific	Pathogen – benefits Host – harmed	Influenza
Mutualism - interaction between two species living in close association where both species benefit	Interspecific	Both benefit	Lichen - an association between an alga and a fungus
Commensalism - interaction between two species living in close association where one species benefits but the other species is neither benefitted or harmed	Interspecific	One benefits, the other neither benefits or is harmed	Clown fish and sea anemones

10 For a farmland area with large rabbit populations, give examples of competition for abiotic resources and biotic resources. What other interactions might also exist in the area?

11 Considering human populations what competitive interactions are evident in your school community?

12 What is allelopathy and how does it increase a plants chance of survival?

13 What is the distinguishing feature between Predator/Prey and Parasite/Host relationships?

14 What factors affect numbers in predator and prey populations?

15 Sometimes interactions between organisms are symbiotic. What does this mean?

16 Complete the following table.

Type of symbiosis	**Who benefits/loses**	**Example**
mutualism		
commensalism		
parasitism/infectious disease		
amensalism		

17 Zooxanthellae are tiny photosynthetic organisms that live in the tissue or the coral polyps (living part of coral – animal). What sort of interaction is involved? Describe the harm/benefit for the coral polyp and the Zooxanthellae.

18 The remora, a sucker-fish, is often found hitching a ride with larger fish such as sharks. What sort of relationship exists? Justify your answer.

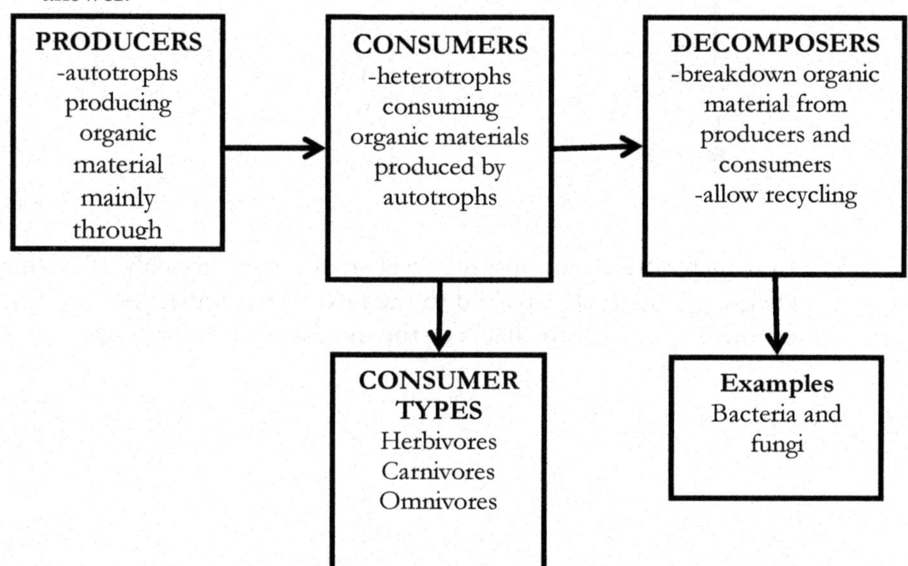

19 What is meant by 'ecological niche'?

20 Explain the disadvantage of niche overlap for two species that live in the same habitat.

21 Which ecological group contains organisms capable of making their own organic materials? Why are they so important in all communities?

22 Describe the diet of the different consumer types giving examples with which you are familiar to illustrate your answer.

23 Why are decomposers so important in all communities even though they are not always an obvious part of the community?

24 In ecosystems, there is a big difference between what happens to energy and what happens to matter over time. Explain what this difference is.

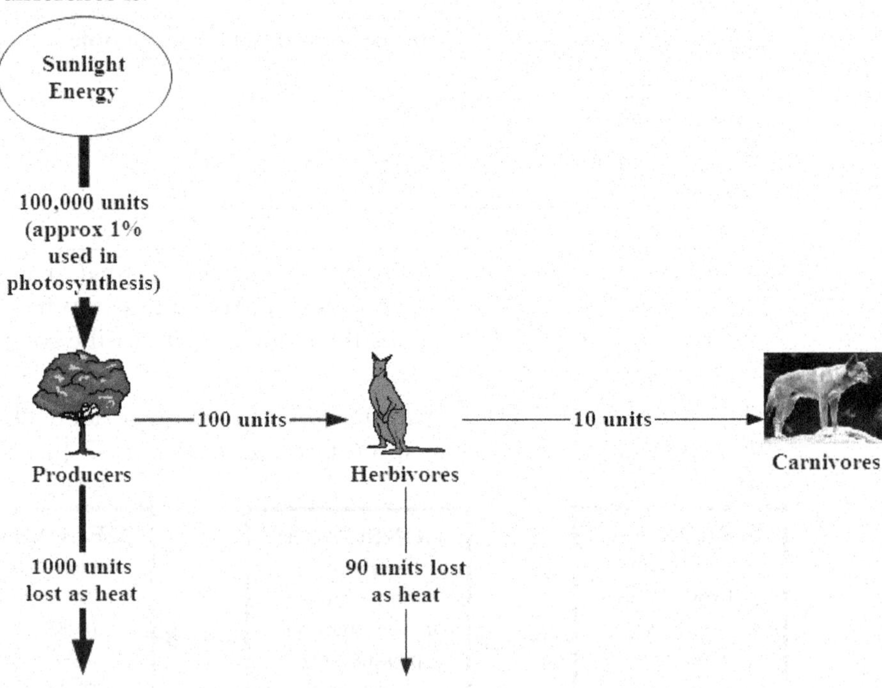

25 In transfer from one trophic level to the next, roughly 10% of the energy at one level, is passed to the next. What impact does this have on the length of food chains or the number of trophic levels?

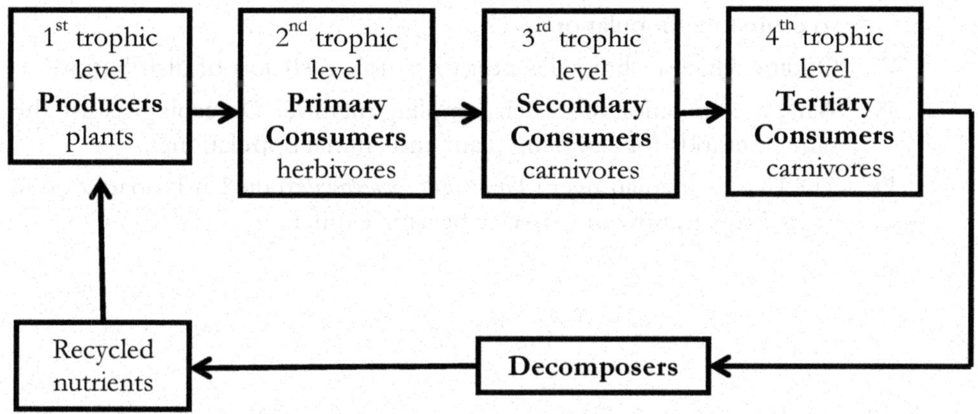

26 In food chains and food webs, what do the arrows represent and which way do they point?

27 Draw a food chain for the following: cow, grass, and a non-vegetarian human.

28 Draw a likely food web for the following group of organisms: grass, small shrubs, human being, fox, rabbit, sheep, wedge-tailed eagle, wild dogs and mushrooms.

29 Using the information in the previous question, predict what would happen to the populations in the short-term, if a disease killed all wedge-tailed eagles.

30 What is the usual short-term consequence on an ecosystem of species competing for resources?

31 What is the usual long-term consequence on an ecosystem of species competing for resources?

32 Distinguish between the terms 'distribution of a species' and 'abundance of a species'.

33 What are the three types of population distribution illustrated in the figure below?

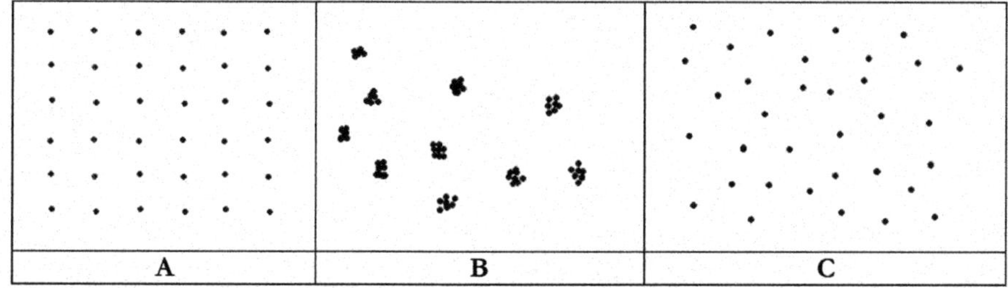

34 B is the most common sort of distribution found in nature. Suggest a reason for this.

35 Sampling techniques are often used to estimate population size. When are they used?

36 What are the three major methods used for sampling populations and so estimating population size?

37 Outline which technique is best to sample each sort of distribution?

38 What are limitations for each sampling method? Do ecologists use the same methods for sampling plant and animal populations?

39 The last Tasmanian tiger (*Thylacinus cynocephalus*) died at Hobart Zoo in 1936. Explain how this species became extinct.

Evolution of Australian Biota Summary

Millions of years ago		Climate	Fossil finds	Plants	Animals
150 140	Gondwana – South America/Africa/ Madagascar/India/Antarctica/Aust/NZ Africa and South America separated India moving north			Conifers, cycads, ferns	dinosaurs
120					
			Lightning Ridge	Pines, ferns and first flowers	First monotremes/birds *Steropodion galmani*
100					
80	Madagascar split from Africa New Zealand split from Australia				
		Cool and wet			
60	South America and Australia/Antarctic separate				Dinosaurs extinct
			Murgon	Rainforests	Oldest marsupial and placental fossils *Thylacotinga bartholomaii* *Tingarmarra porterorum* (condylarth)
		Cool and dry			
40	Australia/New Guinea and Antarctic separate				
			Riversleigh		Modern marsupials *Obdurodon dicksoni* *Thylacinus cynocephalus*
		Warmer			
20			Naracoorte	Grasslands/forests Myrtaceae	Placentals from Asia Megafauna (40 000yrs) *Diprotodon* *Procoptodon* *Thylacoleo*
Present		Warmer and drier		*Eucalyptus*/*Acacia*	Reptiles/marsupials

Past Ecosystems

40 Complete the following summary of evidence of changing environments in Australia.

Evidence of changing Australian environments	Description
Human records	
Geological	
Paleontological	
Living organisms	

41 How are rock paintings dated?

42 What three pieces of information must be known in order to use radiometric dating?

43 What is ice core drilling?

44 What can ice cores indicate?

45 How are gases trapped in ice cores extracted and analysed?

46 Gas analysis of Antarctic ice cores indicate changes have occurred in temperature and atmospheric carbon dioxide. Describe these changes and suggest why they have occurred.

47 After Australia separated from Gondwana, it gradually moved northwards. What were the associated climate changes?

48 Describe the changes in Australia's vegetation as indicated by the fossil record.

49 What is meant by sclerophyll vegetation?

50 Describe the ancestor of the kangaroo and where it lived.

51 Why is the ability to hop an advantage in an arid environment?

52 What are the two major factors that have altered the Australian landscape since European settlement in 1788?

53 What impact did the indigenous people of Australia have on vegetation and wildlife before 1788?

54 A major problem that has been a consequence of changes after settlement has been increasing dry-land salinity. Describe the causes of this problem.

55 Rabbits have been a problem since their introduction in Australia in the 1840s. How have scientists tried to control their population? What is this technique an example of?

Future Ecosystems

56 Why is biodiversity decreasing?

57 Why is it important to maintain biodiversity?

58 How has humans contributed to habitat loss?

59 What is habitat fragmentation?

60 Describe how habitat fragmentation may lead to extinction.

61 List the ways humans have impacted on ecosystems.

62 What is monitoring and how can it be used to manage biodiversity?

63 Define bioindicator.

64 Lichen is used as a bioindicator for air pollution. What are the characteristics of a good bioindicator?

65 How are scientific models used?

66 Explain how species distribution models can be used to predict future impacts on biodiversity.

67 What is the Global Climate Model (GCM)?

68 Summarise how the ecosystems affected by mining sites may be restored by completing the following table.

Process	Description
Clean up of contaminants	
Land form reconstruction	
Soil restoration	
Revegetation	
Fauna recolonisation	

69 List the main causes of land degradation due to agricultural practices.

Section I Questions

70 Organisms live in communities with populations of other species of organisms. Which of the following is not true?

(A) Communities can provide a diversity of habitats for different species.
(B) All of an organism's needs will be met in its habitat.
(C) Some species move from habitat to habitat at different times.
(D) Niche overlap between two species in a community leads to greater competition.

71 In a particular area, biologists will talk about a community of organisms. Communities are made up of:

(A) biomes.
(B) populations.
(C) ecosystems.
(D) abiotic factors.

72 Population size is affected by four things that will determine whether a population grows or shrinks: the interplay of birth rate (B), death rate (D), immigration rate (I) and emigration rate (E).

(A) A population will increase in size if B is greater than D.
(B) A population will decrease in size if D+I is greater than B + E.
(C) A population will decrease in size if D+E is greater than B + I.
(D) A population will increase in size if B+E is greater than D + I.

73 Exponential growth in populations does not normally occur in natural populations because

(A) as environmental resistance increases population growth decreases.
(B) density independent factors prevent population size exceeding the carrying capacity of an area.
(C) density dependent factors such as bushfire and drought become more significant as population size increases.
(D) competition for resources decreases with population increases which prevents excessive population growth.

74 Relationships between organisms in a community can be classified on the benefit or harm to those involved. Organisms where neither side is harmed include

(A) predator/prey relationships.
(B) intra-specific competition.
(C) parasite/host relationships.
(D) mutualism.

Use the information below to answer Questions 75 and 76.

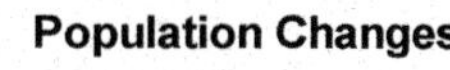

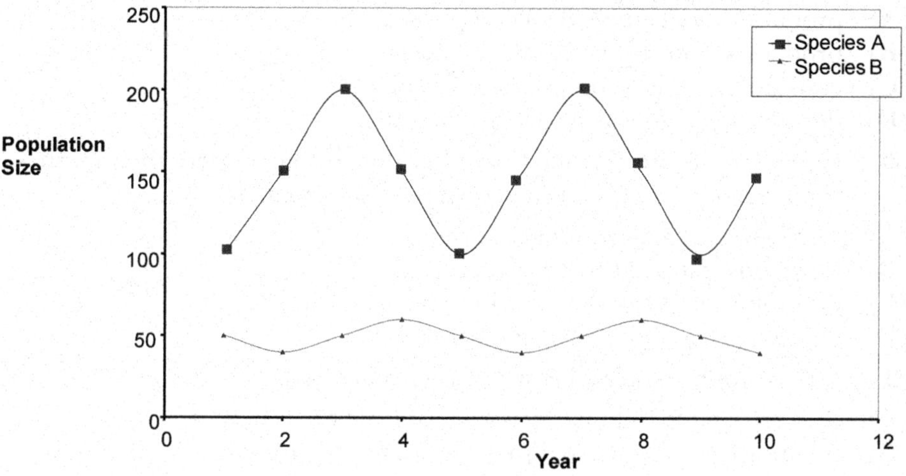

75 Using the information in the graph above, you could conclude the following to be correct.

(A) Species A is the predator of species B.
(B) Species B is the predator of species A.
(C) The predator species will have a number of different prey species.
(D) The peaks in populations of predator and prey species coincide.

76 Using the information in the graph above, you could conclude the following to be correct.

(A) There is a balance between the number of prey and predators.
(B) The number of prey organisms decreases due to disease.
(C) In the 10 years that follow, you could expect the same pattern would continue.
(D) The size of the predator population is independent of the prey population.

77 Parasite/host relationships are often very complex. Parasites have special adaptations to allow them to obtain nourishment from the host. With regard to these relationships, the most accurate statement below is

(A) endoparasites live inside the host.
(B) parasites eventually kill their host.
(C) exoparasites live inside their host.
(D) the host gains benefit from the parasites activities.

78 Mutualism and commensalism are similar in that both members of the relationship

(A) gain benefit.
(B) are harmed.
(C) neither are harmed.
(D) the members can survive independently.

79 Organisms in a community can normally be classified into three categories

(A) producers, consumers & decomposers.
(B) producers, autotrophs & decomposers.
(C) consumers, heterotrophs & autotrophs.
(D) decomposers, autotrophs & producers.

80 Heterotrophs are classified on what they eat and where they are in a food chain. The most complete correct statement is

(A) herbivores eat plant eating animals.
(B) carnivores eat plant material and meat.
(C) decomposers break down dead remains.
(D) omnivores eat herbivores and carnivores.

81 Heterotrophs require digestive systems because

(A) they are unable to build their own organic molecules.
(B) the molecules in the food they eat are too big to pass into the body.
(C) chemical digestion can only occur within cells.
(D) they produce large amounts of waste material.

82 In food chains and food webs, the arrows are best described as presenting

(A) the flow of energy from primary consumers to secondary consumers.
(B) the movement of matter and energy from one trophic level to the next highest level.
(C) movement of matter from consumers to producers and decomposers.
(D) flow of energy from high trophic levels to lower trophic levels.

83 With regards to energy movement through a community, scientists often refer to the '10% rule'. The rule refers to

(A) the fact that only 10% of energy from the sun is harnessed by producers.
(B) the idea that roughly 10% of the energy harnessed at one trophic level is passed onto the next level.
(C) the observation that there are always more producers in a community than there are primary consumers.
(D) that only 10% of the energy stored by producers reaches the top consumer.

84 In a community, energy is passed from one trophic level to the next. Most communities are governed by the 10% rule – that is only 10% of the energy at one level is available to be passed on to the next level. As a consequence of this rule

(A) communities can have up to 10 trophic levels.
(B) a large biomass of producers is usually required to support the rest of the community.
(C) there are more top consumers than primary consumers.
(D) energy can be efficiently recycled within a community.

85 An Australian rural area has the following organisms present: grasses, rabbits, wedge-tailed eagles, mushrooms, rabbits, sheep and foxes. A likely food chain is

(A) wedge-tailed eagle → rabbit → grass.
(B) grass → sheep → fox → toadstool.
(C) grass → rabbit → fox → mushroom.
(D) grass → fox → rabbit → wedge-tailed eagle.

The following information is required for questions ***86–88.***

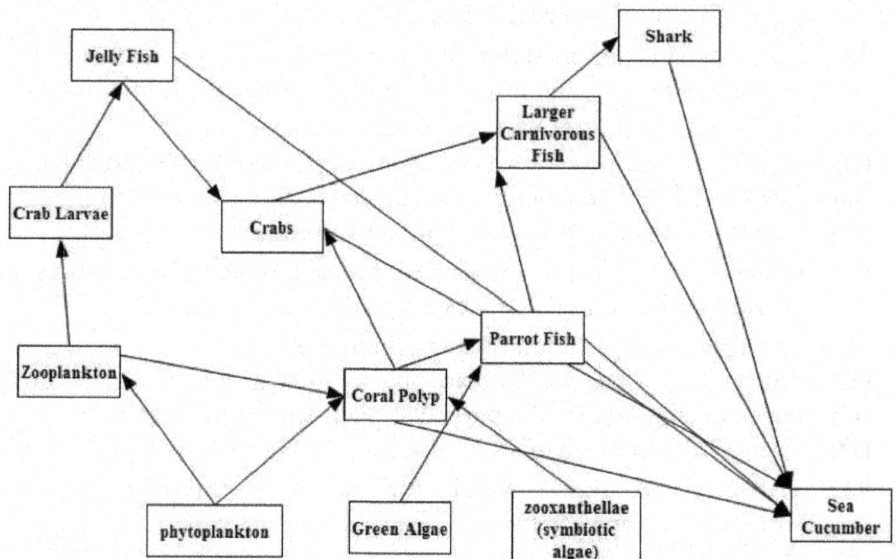

86 The organisms capable of photosynthesis in the food web are:

(A) zooplankton.
(B) coral polyp.
(C) zooxanthellae.
(D) crab larvae.

87 First order consumers in the food web are:

(A) phytoplankton.
(B) zooplankton.
(C) crab larvae.
(D) parrot fish.

88 The longest food chain in the web contains

(A) 4 organisms.
(B) 7 organisms.
(C) 6 organisms.
(D) 5 organisms.

89 Matter and energy flow through ecosystems. A major difference between energy and matter in ecosystems is

(A) energy is recycled in ecosystems.
(B) matter and energy are not recycled in ecosystems.
(C) energy loss from ecosystems is constant over time whereas matter loss is increasing.
(D) matter is recycled whereas as an input of energy is required to replace energy lost from an ecosystem.

90 Scientists studying in different environments need to gather data to measure physical conditions. Which of the following is incorrect?

(A) Air pressure can be measured with a barometer.
(B) Water salinity can be measured by looking at electrical conductivity.
(C) Light intensity differences can be detected with a light meter.
(D) Wind speed can be measured with a hygrometer.

91 Different methods are appropriate for determining the numbers and distributions of different species. Quadrats are useful for

(A) measuring changes in numbers of animals in area.
(B) distribution of a particular plant species in an area.
(C) the absolute number of a particular plant species in an area.
(D) a way of detecting changes in physical conditions in an area.

92 Transects are very useful for providing information about communities for all but one of the following:

(A) identifying community changes in plant communities across a mountain range.
(B) looking at changes in sand dune communities.
(C) looking at the zonation in mussels and barnacles across a rock platform.
(D) determining the change in population density of plants in a pasture.

93 The capture-recapture method relies on which of the following factors:

(A) marking needs to be obvious to allow scientists to find marked individuals.
(B) some individuals are more likely to be captured than others.
(C) predation of individuals is not changed by marking technique.
(D) in the time between samples no marked individuals have died.

94 The best technique to look at changes in distribution of plant life across a sand dune would be

(A) random sampling.
(B) a series of random quadrats.
(C) a series of evenly spaced quadrats.
(D) a transect.

95 In using the 'capture–recapture' method, one of the following is not an assumption.

(A) Marking makes the animal more susceptible to predation.
(B) All individuals have the same probability of being captured and recaptured.
(C) There is little movement of the target species in and out of the area.
(D) The marked animals will survive long enough to be recaptured.

96 The extinction of the Tasmanian tiger was due to

(A) hunting by humans.
(B) habitat destruction.
(C) disease.
(D) all of the above.

97 It is difficult to date rock paintings because

(A) the paintings have been exposed to wind.
(B) the rock on which they are painted contains little carbon-14.
(C) the paintings have been exposed to water.
(D) little organic material remains.

98 The remains of an old wasp nest covering a rock painting was dated using carbon-14 and found to be 17 000 years old. This suggests the paintings are

(A) younger than 17 000 years old.
(B) 17 000 years old.
(C) older than 17 000 years old.
(D) at least 50 000 years old.

99 The layers seen in ice cores are formed from

(A) compressed sediments.
(B) compressed snow.
(C) compressed seawater.
(D) compressed freshwater.

100 Ice cores indicate

(A) only the biotic factors present at the time the layers were formed.
(B) only the abiotic factors present at the time the layers were formed.
(C) both biotic and abiotic factors present at the time the layers were formed.
(D) conditions in the Antarctica have changed very little throughout time.

101 As Australia broke away from Gondwana and moved north, the climate became

(A) drier and hotter.
(B) drier and colder.
(C) wetter and hotter.
(D) wetter and colder.

102 Which of the following adaptations is **not** an adaptation for living in a hot, dry climate?

(A) ability to sweat
(B) ability to hop
(C) presence of hard, small leaves
(D) ability to produce concentrated urine

103 Decreased biodiversity

(A) decreases the chances of extinction
(B) increases the chances of extinction
(C) can increase or decrease the chance of extinction
(D) does not affect the chance of extinction

104 Organisms that are good bioindicators

(A) do not react to change in the same way other organisms in its ecosystem do.
(B) react to change in different ways.
(C) are sensitive to change and are easy to observe and sample.
(D) do not react to changes in their environment.

105 Scientific modelling allows scientists to

(A) predict what might happen to biodiversity in the future.
(B) change a variable and then directly measure the effect this has on a population.
(C) directly measure the effect of changing many variables at once.
(D) prove that changing a variable will produce a certain result.

Section II Questions

106

Giving examples, explain the difference between the following different pairs of terms.

(a) populations and communities [2 marks]
(b) biotic factors and abiotic factors [2 marks]

[Total 4 marks]

107

The following table lists a number of interactions between species. For each, name the relationship and give a reason for your choice. The first example is completed.

	Relationship	Reason
tapeworm in pigs	parasite/host	tapeworm benefits, pig is harmed
sheep and kangaroos in a paddock		
lichen – consisting of a fungus and algal cells		
sea anemone catching small fish		
mosquito on you!		
rabbits in a paddock		

[Total 5 marks]

108

(a) Define and contrast the following two pairs of terms.

(i) producer and autotroph [1 mark]

(ii) consumer and heterotroph [1 mark]

(b) Complete the following table adding the most appropriate description of the relationship between the two organisms in the right-hand column.

Organisms	Relationship
2 sheep in a paddock	
A sheep and a rabbit in a paddock	
Koala and a gum tree	
Fungi and algal components of a lichen	
Flea and a dog	
Mould and some old bread	

[5 marks]

Commensalism refers to a relationship between two organisms in which one gains a benefit and the other does not appear to gain a benefit or be harmed.

(c) Give an example of commensalism. [1 mark]

(d) For your example, do you think the definition is absolutely correct? Discuss. [2 marks]

[Total 10 marks]

109

'No man is an island, entire of itself; every man is a piece of the continent, a part of the main; if a clod be washed away by the sea, Europe is the less ... any man's death diminishes me, because I am involved in mankind ...'
John Donne poet (1572–1631)

This quotation could be rewritten in terms of the organisms with an ecosystem. An ecosystem consists of producers, consumers and decomposers. The activities of organisms at each of the different trophic levels have impacts at other levels.

(a) Suggest two things that could happen in a community if the major producer organism is removed. [2 marks]

(b) In a community that includes at least, grass, rabbits, feral cats and wedge-tailed eagles, what might happen if the top predator – the eagle – is removed? (Note that the eagle eats feral cats and rabbits.) [2 marks]

(c) Introduction of myxomatosis and the calicivirus had a dramatic effect on rabbit populations. Describe the probable impact on other populations in the area. [2 marks]

(d) In agricultural areas of Australia, suggest three major changes that have been bought about since European settlement. [3 marks]

(e) Many scientists consider that the increased greenhouse effect has been bought about by industrial nations accelerated burning of fossil fuels. Considering the quotation at the start of the question, if there are solutions how would they have to be developed and successfully applied. [2 marks]

[11 marks]

110

A group of students studying the saltmarsh plant *Salicornia*, used a 1 m^2 quadrat to obtain the following results.

Quadrat	**Number of *Salicornia* plants**
1	16
2	3
3	14
4	15
5	2
Total	

(a) Calculate the *Salicornia* abundance per square metre. [1 mark]

(b) Estimate how many *Salicornia* plants would be found in 80 m^2. [1 mark]

(c) Assess the validity of the students' results. [2 marks]

(d) Would using quadrats be an accurate method to use to estimate the population size of ducks? Justify your answer. [2 marks]

[Total 6 marks]

111

In a pigmy-possum population study in an alpine region, 20 pigmy-possums were captured, tagged and released. One month later scientists captured 10 possums from the same area. Of these, four were tagged.

(a) Estimate the pigmy-possum population size in the study area. [2 marks]

(b) Assess the validity of the results of this sampling. [2 marks]

[Total 4 marks]

112

(a) Briefly describe the dominant vegetation present when Australia was part of Gondwana. [2 marks]

(b) What changes in vegetation occurred after Australia separated from Gondwana? [2 marks]

(c) What is thought to have brought about these changes in vegetation? [2 marks]

[Total 6 marks]

113

Consider the cross-section through the Australian Great Dividing Ranges from the east cost to the dry inland.

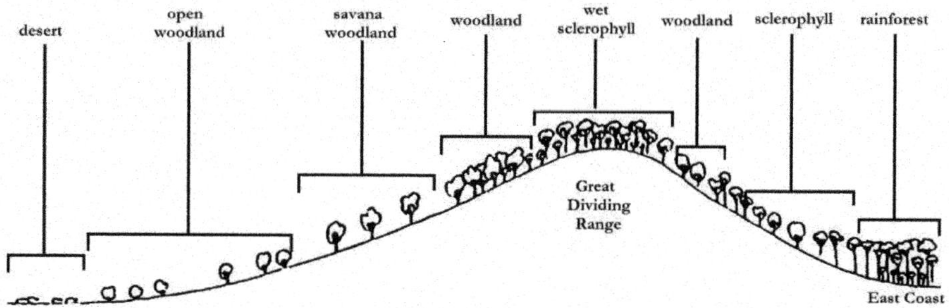

(a) Complete the sentence: In terms of techniques to examine communities, the above diagram is an illustration of a very large __________________________. [1 mark]

(b) Describe two major climatic physical changes that occur from east to west that help to explain the changes in vegetation. [2 marks]

(c) Describe two major differences you would notice sitting in a rainforest compared to sitting in open woodland. [2 marks]

In Northern Queensland, a great deal of the coastal rainforest has been cleared to allow for sugar cane farming and cattle grazing. Clearing occurred up to the edge of rivers and creeks. Rainfall is very high in some of these areas and erosion has often been severe. This has led to high sediment loads in rivers leading to the coast adjacent to the Great Barrier Reef.

(d) What impact would this have had on the nearby coastal reefs? Explain why this would cause problems. [2 marks]

(e) What changes in clearing practices might have reduced the impact of this problem? [2 marks]

(f) Removal of mangroves (flowering plants that live in salt water) all around coastal Australia has had an impact on water quality and fish resources. Explain the two major problems caused by mangrove removal. [2 marks]

[Total 11 marks]

Chapter 6
Suggested Answers for Chapters 2 to 5

Chapter 2: Cells as the Basis of Life
Content Review Questions – Suggested Answers

2-1 The independent variable is the factor the scientist deliberately changes. The dependent variable is the factor the scientist measures. Controlled variables are factors that are kept the same for both experimental and control groups.

2-2

Variable	Categorical	Numerical
qualitative/quantitative	qualitative	quantitative
description/number	description	number
graph	bar charts/pie graphs	scatter plots/line graphs
examples	- gender of children in a family - birth order	- number of children in a family - height of children in a family

2-3 The results of the experimental control group allow the scientist to be confident that the results obtained by the treatment group are due to the independent variable only and not some other factor.

Note: it is not good enough to say that the control is there as a 'standard of comparison'.

2-4 The experimental control group is a set-up that is used to ensure that the results are due to the independent variable only, whereas controlled variables are the factors that are the same for the experimental control group and the treatment group.

2-5 No. Experimental accuracy refers to how close the result is to the value obtained if the quantity could be measured perfectly, whereas precision refers to how close in value are repeated measurements.

2-6 Experimental precision can be improved by repeating the experiment many times and by practising experimental techniques.

2-7 A large number of individuals in a treatment group minimises the effect on the results of natural variation in the tested organisms.

2-8 Valid results are those obtained from controlled experiments where there is only one independent variable. Reliable results are close results obtained when the experiment is reproduced by other scientists.

2-9 Primary data is data that you collect whereas secondary data is data that someone else collects.

2-10 The cell theory states:

- all living organisms are made up of one or more cells and the products of cells (Matthias Schleiden and Theodor Schwann, 1839)
- cells are the smallest units that show the properties of life (Matthias Schleiden and Theodor Schwann, 1839)
- cells come from pre-existing cells (Rudolf Virchow, 1858).

2-11

	Human eye	**Light microscope**	**Electron microscope**
Magnification	1x	up to 1 500x	1 000 000x
Resolution	Around 200 μm	Around 0.2 μm	Around 0.001 μm
How they work	Light rays bent by the eye's lens forming an image on the retina	Light rays bent by glass lens forming an image in our eye	Electron beams bent by magnets forming an image on a screen
Advantages		Samples prepared quickly; application varied because coloured stains can be used and living cells can be viewed	High magnification and resolution allow objects as small as molecules to be viewed
Disadvantages		Limited visible detail	Only non-living sections can be viewed because electrons must be kept in a vacuum or the beam is scattered. Does not show colour. Cost.

2-12 Magnification is a measure of how many times larger the object appears whereas resolution refers to how much detail can be seen.

2-13 Total magnification is equal to the magnification of the objective lens multiplied by the magnification of the ocular (eyepiece) lens.

2-14 Some structures absorb more stain than others do. This increases the contrast and makes them easier to see. Stains can also be used to locate where specific biomolecules are within the cell. For example, iodine shows the location of starch within the cell.

2-15 Stains kill the cells so living samples cannot be observed. Stains also mask the natural colour of structures.

2-16

Date	**Technological advance**	**Significance**
Mid 1600s	Compound microscope (Robert Hooke)	Cork was seen to be made of small units called 'cells'
Late 1600s	Improved lenses leading to better resolution and magnification (Anton van Leeuwenhoek)	Greater detail could be seen, and more organisms observed supporting the idea that all organisms are made of cells. First to view bacteria cells
1800s	Better stains and improved sectioning techniques	Allowed thinner sections to be prepared and greater detail to be seen.
1930s	Phase contrast microscopes (Fritz Zernike)	Allowed the study of unstained living cells

1930s	Transmission electron microscopes (Ernst Ruska)	Allowed the fine detail of the internal structure of cells to be studied
1930s	Scanning electron microscopes (Max Knoll)	Allowed the 3D surface of cells to be studied
1930s-40s	Cell fractionation (Albert Claude)	Allowed the study of individual components of cell
1940s	Radioactive tracers	Allowed the study of the sequence of metabolic reactions within cells
1950s	Confocal microscopes (Marvin Minsky)	Improved the study of cells by producing sharply focused images

All the previous technological advances have allowed the structure of living organisms to be studied in great details. All observations support the idea that living organisms are composed of one or more cells and the products of cells. Cells are the smallest units that show the properties of life.

2-17 All cells have a plasma membrane, cytoplasm, DNA (at least at some stage) and ribosomes.

2-18

	Type of Cell	
	Prokaryote	**Eukaryote**
Size	Small and lack specialisation	Relatively large
Structural difference	No membrane-bound organelles Circular chromosome	Membrane-bound organelles present Linear chromosomes
Examples	Monera (bacteria and cyanobacteria)	Protists, plants, animals, fungi

2-19 Ribosomes are found within both eukaryotic and prokaryotic cells.

2-20

Organelle/ Cell part	**Structure**	**Function**	**Observed with light/ electron microscope**
Cell wall	Plants – cellulose Bacteria – protein and murein Fungi – chitin	Support	Light microscope
Cell (plasma) membrane	Phospholipid bilayer	Regulates movement in and out of a cell	Electron microscope. No membranes can be seen using the light microscope because they are too thin. The positions of the cell membrane, vacuole membrane and other internal membranes envelopes can often be identified.

Nucleus	Membrane-bound organelle containing chromosomes (DNA)	Controls the activities of the cell	Light microscope
Nucleolus	Area within the nucleus Contains RNA	Involved in ribosome formation	Light microscope
Nuclear membrane	Phospholipid bilayer	Regulates movement in and out of the nucleus	Electron microscope
Cytoplasm	Contents of the cell Water, soluble materials and organelles.	Site of cellular activities	
Cytosol	The fluid component of the cytoplasm	Site of cellular activities	
Ribosomes	RNA	Site of protein synthesis	Electron microscope
Golgi body (apparatus)	Phospholipid bilayer Stack of membrane-bound sacs	Site of packaging and storing of chemicals to be secreted	Light microscope with staining
Mitochondria	Phospholipid bilayer Also contain DNA and ribosomes	Site of aerobic respiration	Light microscope with staining
Plastids Chloroplasts Leucoplasts Chromoplasts	Phospholipid bilayer Grana and stroma Also contain chlorophyll, DNA and ribosomes Membrane sacs containing starch Membrane sacs containing pigments	 Involved in photosynthesis Store starch Synthesise and store pigments	Light microscope
Centrioles	Microtubules Protein	Involved in cell division in animals	Electron microscope
Chromosomes	DNA	Contain genetic information	Light microscope

Vacuoles	Phospholipid bilayer Membrane sacs	Store water and mineral ions	Light microscope
Endoplasmic reticulum - rough ER	Phospholipid bilayer Membranes with ribosomes (RNA)	Intracellular transport Protein synthesis	Electron microscope
- smooth ER	phospholipid bilayer membranes	Produces lipids Intracellular transport	Electron microscope
Vesicles	phospholipid bilayer membrane sacs	Transport cell products to the cell membrane	Electron microscope
Lysosomes	Phospholipid bilayer Membrane sacs containing digestive enzymes	Destroys old cells and organelles	Electron microscope
Microtubules and microfilaments	Protein	Give rigidity to the cell and aid in movement of some organelles	Electron microscope
Cilia and flagella	Microtubules Protein	Involved in locomotion	Light microscope

2-21 The presence of membrane-bound organelles. If present then the cell is eukaryotic, as prokaryotes do not have membrane-bound organelles.

2-22 The presence of a cellulose cell wall. If present, the cell is from a plant, as animal cells do not have a cell wall. Note, if plastids (e.g. chloroplasts) are present then the cell is from a plant but not all plant cells have plastids, so this answer is not always true. Many plant cells have a large vacuole but not all.)

2-23 Plant cells have a cellulose cell wall whereas the cell wall of prokaryotes is made of peptidoglycan. Plant chromosomes are linear whereas prokaryotic chromosomes are circular. Plants cells contain membrane-bound organelles whereas prokaryote cells do not.

2-24 Chloroplasts are large membrane-bound organelles that can be seen with a light microscope. Within the outer membrane are a series of folded membranes that from stacks called grana. The grana are green as they contain chlorophyll, the pigment that absorbs red and blue light. The grana are the site of the light dependent reactions. The spaces between the grana are known as stroma and are the site of the light independent reactions.

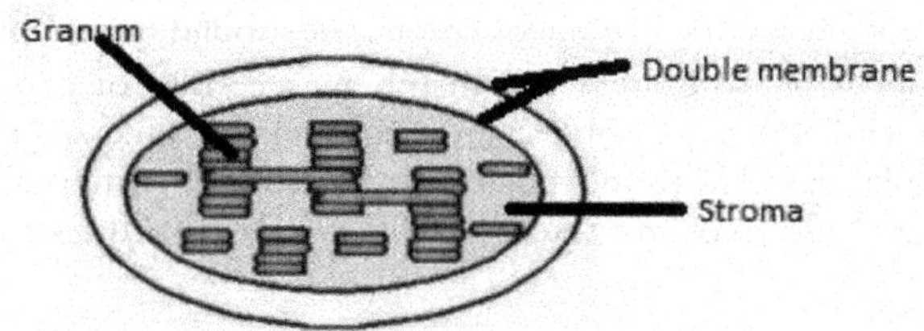

2-25 Mitochondria are large membrane-bound organelles found in the cytoplasm. They can be seen with a light microscope if stained but structural detail can only be seen with an electron microscope. Within the outer membrane is a highly folded internal membrane. The projections formed from these folds are called cristae.

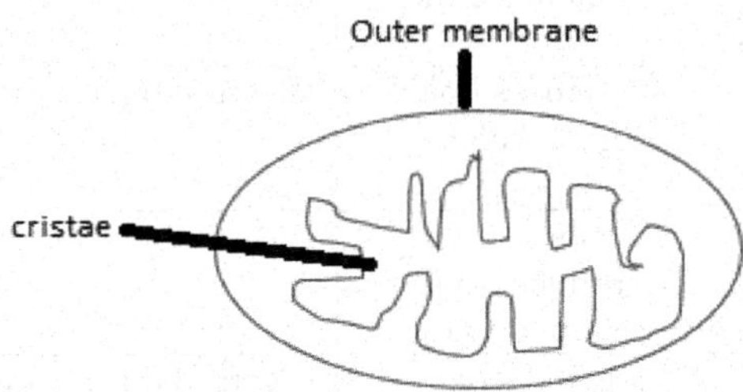

2-26 Mitochondria and chloroplasts are similar in structure to some current free-living bacteria. It is suggested that some large bacteria ingested whole smaller bacteria and this lead to the evolution of eukaryotic cells.
A double membrane surrounds both mitochondria and chloroplasts, and both contain DNA and RNA.

2-27 Six functions of cell membranes are:

- control of substances moving into and of cells
- compartmentalisation of internal cellular spaces – organelles
- intercellular communication
- cell-to-cell recognition
- site of many cellular reactions
- electrical activity in excitable cells.

2-28 Selectively permeable means that some molecules can pass through the membrane while others cannot.

2-29

	Structure	Function
A	Carbohydrate	Cell recognition
B	Protein channel	Movement of ions Note: aquaporins are a type of protein channel that allow water and urea to diffuse through the membrane.
C	Protein carrier molecule	Site of facilitated diffusion and active transport
D	Cholesterol	Stability of the membrane
E	Phospholipid bilayer	Major structural component Site where lipid soluble and small uncharged molecules are exchanged

2-30 Phospholipids and proteins are the two major compounds found in membranes.

2-31 The fluid mosaic model proposes that cell membranes consist of a bi-layer of phospholipids with embedded proteins. The membrane is fluid as individual phospholipid molecules are able to move laterally. Fat soluble materials can pass through the lipid part of the membrane. Other materials pass through protein channels.

2-32 Four factors that increase the rate of diffusion are:
- concentration gradient. The greater the difference in concentration, the greater the rate.
- temperature. The higher the temperature, the greater the rate.
- size of molecules. The smaller the molecules, the greater the rate.
- state. Diffusion occurs faster through a gas than through a liquid.

2-33 Diffusion gradient refers to the difference in concentration of a substance from one area to another. The greater the difference, the greater the diffusion gradient.

2-34 When cells are placed in distilled water, there is a net movement of free water into the cells by osmosis as there is a greater concentration of solutes inside the cell than outside. As a result, the cells swell. The plant cell has a cell-wall, and this exerts a pressure that stops the plant cell from exploding. Animal cells do not have a cell wall and, so they explode.

2-35

	Simple Diffusion	**Channel-mediated**	**Carrier-mediated**		**Vesicle-mediated (endocytosis /exocytosis**
			Facilitated diffusion	**Active transport**	
Where	Lipid bi-layer Transient pores	Protein channels	Carrier proteins		Large areas of membrane
What	Lipid soluble molecules and small uncharged molecules	Usually ions (aquaporins – water and urea)	Sugars, amino acids and ions	Anything against the conc. gradient	(Large) Molecules required in large amounts
Direction	High concentration to low concentration			Low conc. to high conc.	
Energy	passive			active	

2-36

	Definition	**Comparison**
Diffusion	Process where molecules of a substance move from a region of high concentration to a region of lower concentration resulting in an even spread of molecules. Can occur in gases or liquids.	No external energy is required. Movement is due to the random movement of all molecules. Note, however, that the rate of diffusion increases with temperature. Movement of the molecules is down the concentration gradient.
Osmosis	A special case of diffusion – refers to the **net** movement of free **water** molecules across a **semi-permeable** membrane from an area of **low solute** concentration to an area of **high solute** concentration.	As above, but note that the movement of molecules refers to water only. Water molecules are moving down their concentration gradient.

Facilitated diffusion	Case of diffusion where movement of materials is aided (facilitated) by special protein carrier molecules. Molecules move down their concentration gradient.	Increases rate of movement of materials across membranes when the materials concerned do not dissolve readily in the lipid layer of membranes.
Active transport	Protein carrier molecules involved in moving materials across membranes **against a concentration gradient** – that is, the material moved to a region where it is in higher concentration.	Energy is required. This process can only occur in living cells that are respiring and so producing ATP. Process is very important in plant root hairs where ions are actively absorbed from the soil and in the mammalian digestive system.
Vesicle-mediated transport	The cell membrane and vesicles are involved in the movement of large molecules into or out of the cell.	Energy is required. Involves the movement of very large molecules and vesicles fusing with the cell membrane.

2-37 Size and polarity determine where and how hydrophobic and hydrophilic molecules cross membranes. Hydrophobic and small-uncharged molecules diffuse across the lipid bilayer, ions and water through protein channels, sugars, amino acids through carrier proteins, any substance against the concentration gradient through carrier proteins and large molecules required in large amounts via vesicles.

2-38 Exocytosis and endocytosis are types of vesicle-mediated transport. Exocytosis is the bulk movement of molecules produced by the cell (e.g. hormones and neurotransmitters), out of the cell. Endocytosis is the bulk movement of molecules into the cell. Pinocytosis and phagocytosis are both types of endocytosis. Pinocytosis is the bulk movement of liquids (e.g. fat droplets) into the cell and phagocytosis is the bulk movement of solids (e.g. bacteria) into the cell.

2-39 SA:V is a measure of the area of the cell surface compared with the volume of its contents.

2-40 Cells must obtain requirements and remove wastes at a rate that allows the chemical reactions required for life to occur. As the SA:V of the cell is determined by its size, the size of the cell will determine how fast diffusion occurs. If a cell is too large the rate of diffusion will be too slow to maintain supply and removal of wastes.

2-41 A rectangular box.

2-42 Cells are generally microscopic so that they have a large SA:V ratio to allow efficient movement of materials into and out of the cell. This includes nutrients such as glucose, oxygen and removal of wastes such as carbon dioxide.

2-43 Cells that tend to be flattened or have long extensions and irregular surfaces will have a greater SA:V ratio.

2-44 Even though nerve cells may be metres long they are microscopic in diameter and will therefore have a high SA:V ratio.

2-45 An organic compound is a complex carbon-containing molecule produced by living organisms. They usually also contain hydrogen and oxygen.

2-46 Other materials that do not contain carbon and/or are from non-living sources are inorganic compounds. Examples are: salts and vitamins.

2-47 From a study of physics, energy is the potential to do work. In biology that work describes the chemical reactions and the results of those chemical reactions that occur in cells of an organism. For example, energy is needed to do the work of synthesising (making) new materials in cells and allowing contraction in muscle cells.

2-48 The sun is the main source of energy for life on Earth.

2-49 Autotrophs through photosynthesis, transform light energy into chemical energy stored in organic compounds. The energy stored in organic compounds can later be transformed into forms that allow cells to function and live. Heterotrophs cannot photosynthesise therefore they rely on autotrophs for the initial energy transformation.

2-50

Plants use	To make
Nitrates	Amino acids Nucleic acids
Phosphates	ATP Phospholipids Nucleic acids

2-51 Heterotrophs cannot produce their own organic materials and rely on production of organic material by autotrophs directly in the case of herbivores (plant-eaters) or indirectly in the case of carnivores.

2-52

$$6CO_2 + 12H_2O \xrightarrow[\text{chlorophyll}]{\text{light}} C_6H_{12}O_6 \quad + \quad 6O_2 \quad + \quad 6H_2O$$

(Note that you can cancel 6 molecules of water from both sides of the equation and be mathematically correct. From a biochemical point of view, water is both a product and a reactant in the photosynthesis equation.)

2-53 Factors that affect the rate of photosynthesis:
- amount of light
- frequency of light
- presence of chlorophyll
- presence of water
- presence of carbon dioxide
- temperature.

2-54 $C_6H_{12}O_6 + 6O_2 + 36\ P_i + 36\ ADP \rightarrow 6CO_2 + 6H_2O + 36\ ATP$

2-55 The energy stored in the bonds of ATP is the immediate source of energy for cells. The energy is released as ATP is broken down to ADP and phosphate.

2-56 The energy released allows the cell to grow, divide, repair, move, synthesise new molecules, active transport, signal and maintain cell structure.

2-57 Heat is also released during respiration and is lost to the surroundings.

2-58 Factors that affect the rate of cellular respiration:
- presence of oxygen
- presence of glucose
- temperature.

2-59

Biochemical process	**Input**	**Output**
Photosynthesis	Carbon dioxide Water Light	Glucose Oxygen Some water
Aerobic respiration	Glucose Oxygen ADP P_i	ATP (36) Carbon dioxide Water Heat
Anaerobic respiration in plants	Glucose ADP P_i	Alcohol Carbon dioxide ATP (2) Heat
Anaerobic respiration in animals	Glucose ADP P_i	Lactic acid ATP (2) Heat

2-60 Differences between aerobic respiration and anaerobic respiration:
- aerobic respiration has different reactants. In aerobic respiration oxygen is present whereas in anaerobic respiration oxygen is not present.
- the products produced are also different. In aerobic respiration glucose is broken down to carbon dioxide molecules that are very small whereas in anaerobic respiration lactic acid/alcohol are produced and these are still large molecules.
- less ATP is produced in anaerobic respiration than in aerobic respiration.
- chloroplasts and mitochondria both contain DNA and RNA. The structures of these are very similar to DNA and RNA found in prokaryotes.

2-61

<table>
<tr><th></th><th>Autotrophs</th><th>Heterotrophs</th></tr>
<tr><td>Energy source</td><td>sun</td><td>eating autotrophs and other heterotrophs</td></tr>
<tr><td rowspan="2">CO_2</td><td>required for photosynthesis</td><td></td></tr>
<tr><td colspan="2">produced in respiration</td></tr>
<tr><td rowspan="2">O_2</td><td>produced in photosynthesis</td><td></td></tr>
<tr><td colspan="2">required for respiration</td></tr>
<tr><td rowspan="2">H_2O</td><td>required for photosynthesis</td><td></td></tr>
<tr><td colspan="2">required for structure, reactions and exchange of materials
produced in cellular respiration</td></tr>
<tr><td>N</td><td colspan="2">required to add to the carbohydrate produced in PHS to make proteins and nucleic acids</td></tr>
<tr><td>P</td><td colspan="2">required to add to the carbohydrate produced in PHS to make nucleic acids</td></tr>
<tr><td>S</td><td colspan="2">required to add to the carbohydrate produced in PHS to make proteins</td></tr>
</table>

2-62 Plants respire all the time and photosynthesise in the light. Oxygen uptake is a measure of aerobic respiration as oxygen is a reactant. Oxygen is produced in photosynthesis therefore total oxygen exchange is a measure of net photosynthetic rate.

2-63 Carbon dioxide uptake and output could be used to measure net photosynthetic rate.

2-64

	Material	Where they cross the cell membrane	Process
	Oxygen	Lipid bilayer	Diffusion
	Water	Protein channels (aquaporins) A little across the lipid bilayer	Osmosis
	Glucose	Carrier protein	Facilitated diffusion
	Carbon dioxide	Lipid bilayer	Diffusion
	Urea	Protein channels (urea transporters) A little across the lipid bilayer	Diffusion

2-65 Catalysts speed up chemical reactions as do enzymes. Enzymes are organic as they are proteins that are produced by living organisms.

2-66 'Rate of reaction' refers to the amount of product produced in a given time. For example, the rate of reaction could be measured in micrograms of product produced per minute.

2-67 Without enzymes, the chemical reactions occurring inside cells would occur too slowly to sustain life.

2-68 Names of enzymes usually end in 'ase'.

2-69

	Intracellular enzyme	Extracellular enzyme
Site of production	Within cells	Within cells
Site where reaction is catalysed	Within cells	Outside cells
Example	Catalase	Lipase

2-70 The reactants in an enzyme-catalysed reaction are called substrates.

2-71 Enzymes are proteins. The folding of the protein into its tertiary structure provides a site that matches with the shape of specific substrates. This site is called the active site. Only the correct enzyme with the correct molecular shape will bind to the substrate molecule, in the same way that a particular key will open a particular lock. The result is that the active site means that enzymes are specific to particular substrates and therefore they can only speed up specific reactions.

2-72 The 'lock and key' model suggests that there is a perfect fit between the shape of the substrate and the active site of the enzyme whereas the 'induced-fit' model suggests that the fit is not perfect.

2-73 enzyme + substrate → enzyme-substrate complex →enzyme + product

2-74 Enzymes are not destroyed or ultimately altered by the reactions they catalyse. This means they can be reused.

2-75 Enzymes do not increase the amount of product produced. They alter the time it takes to produce the product.

2-76 Enzymes are proteins and therefore they can be denatured by:
- high temperatures (break H bonds and van der Waals forces destroying the quaternary and tertiary structure)
- extremes of pH (break ionic bonds, destroying the quaternary and tertiary structure).

2-77 Denaturation by extremes of pH are usually reversible but denaturation by high temperatures, i.e. about 60°C is not. This means the active sites are permanently altered and the enzyme can no longer function. In terms of the lock-and-key hypothesis, the shape of the key is altered.

At low temperatures, the number of collisions between the substrate and enzyme are few so the reaction rate is low. As the temperature increases, the number of collisions increases, resulting in increased reaction rate. This continues until the temperature reaches a point where the structure of the enzyme begins to change shape (denature). The reaction rate then decreases.

2-78 Factors that affect enzyme-catalysed reactions are:
- temperature
- pH
- concentration of substrate
- concentration of enzyme
- concentration of product.

2-79 Aerobic respiration and photosynthesis involve many sequenced reactions. Each reaction is catalysed by an enzyme. The membranes form compartments that keep the substrates and enzymes together thus increasing the rate and efficiency of the overall reactions.

Section I Questions

2-80 The length of Organism Y is
(C) 60 µm. (Organism Y is about 2 divisions of the scale in length therefore 2×30 µm.)

2-81 All cells have
(C) ribosomes. (Only eukaryotic cells have nuclei, vacuoles and mitochondria.)

2-82 Which of the following organelles would be found in the cells of a prokaryote?
(A) ribosomes. (Prokaryotes do not have membrane-bound organelles. All other alternatives are membrane-bound.)

2-83 Prokaryotic cells
(C) contain circular chromosomes. (Prokaryotes are able to photosynthesise as they have chlorophyll in the cytoplasm. They are smaller than eukaryotes and reproduce sexually by conjugation.)

2-84 All bacteria are
(C) prokaryotes. (All bacteria are prokaryotic. Some are harmful, decomposers and photosynthetic.)

2-85 In eukaryotic cells
(A) the rough endoplasmic reticulum is the site of protein synthesis. (The Golgi body is the site where material is packaged for export out of the cell, mitochondria are the site of aerobic respiration and lysosomes contain digestive enzymes.)

2-86 For each cell, it would be reasonable to expect the student to be able to see
(D) a nucleus. (Chromosomes would only be seen in a dividing cell, chloroplasts in a photosynthesising cell, and vacuoles in animal cells are small.)

2-87 A structure not found in all living plant cells is a

(C) chloroplast. (Chloroplasts are only found in photosynthetic cells. Not all plant cells are photosynthetic, for example, root cells. The other structures are found in all living plant cells.)

2-88 In plant and animal cells

(A) the rough endoplasmic reticulum is the site of protein synthesis. (The Golgi body is the site where material is packaged for export out of the cell, mitochondria are the site of aerobic respiration and lysosomes contain digestive enzymes.)

2-89 Membranes

(A) control what enters and leaves a cell. (Membranes are composed of a lipid bilayer, they are fluid in structure and are selectively permeable.)

2-90 After 4 hours

(C) the artificial cell will have decreased in size. (Water will have moved out of the artificial cell by osmosis. Sucrose would not move into the cell, as the membrane is not permeable to sucrose.)

2-91 The changes observed are due to

(C) osmosis. (See previous explanation.)

2-92 Active transport

(B) occurs through protein carrier molecules in the membrane. (Active transport requires energy for movement against the concentration gradient.)

2-93 When comparing diffusion and active transport

(D) in active transport molecules move from where they are in low concentration to where they are in high concentration whereas in diffusion the molecules move from where they are in high concentration to where they are in low concentration. (Fact.)

2-94 Cells are small because this provides

(B) a large surface area compared with volume for exchange of molecules with the environment. (Most molecules are exchanged by diffusion. This requires a high SA:V for a rate of exchange that will sustain life.)

2-95 Which of the cells has the lowest Surface Area to Volume ratio?

(A) cell K. (Spheres have a low SA:V compared to other shapes.)

2-96 It is likely that cell M

(A) is a site of active uptake of mineral ions. (You are told cell M has many mitochondria. Mitochondria are the site of aerobic respiration therefore the cell produces a large amount of energy. Active transport requires energy.)

2-97 With respect to organic molecules,

(A) they contain carbon. (Organic molecules can be broken down and are made by all living organisms. Water is an inorganic molecule.)

2-98 Organic molecules contain

(B) carbon, hydrogen and oxygen. (Ozone is not an element, iron is not found in all organic molecules, but all organic compounds contain carbon.)

2-99 Inputs of photosynthesis are

(A) carbon dioxide, water and light. (Oxygen is an output.)

2-100 When green plants photosynthesise, they produce complex organic substances. In this process,

(A) light energy is converted into and stored as chemical energy. (Fact – other alternatives are totally incorrect. Note that heat is the least useful form of energy for organisms.)

2-101 In photosynthesis, producers convert light energy to chemical energy. One of the following is not directly necessary for photosynthesis.

(D) oxygen (Oxygen is a product of photosynthesis – not a reactant)

2-102 One of the following is not necessary for aerobic cellular respiration

(A) carbon dioxide. (This is a product of respiration. Glucose and oxygen are reactants for aerobic reparation. All reactions in the body occur in solution – therefore water is present.)

2-103 Plants and animals require

(A) oxygen. (Both plants and animals respire, and oxygen is required for aerobic respiration.)

2-104 Phototrophs

(B) convert the energy in sunlight into the energy stored in the bonds of organic molecules. (Phototrophs only photosynthesise in the light. At low light intensities and in the dark, they take up oxygen and release carbon dioxide.)

2-105 Light energy is the energy source for most ecosystems. Light energy is used by autotrophs to produce organic material. The energy has been converted to chemical energy that can be stored. When compounds are broken down some energy is unavoidably lost as

(C) heat energy. (Fact – some heat is lost in all energy transfers in living systems)

106 Enzymes

(D) are reused. (Enzymes increase the rate of reactions that would occur anyway. They are not steroids and are not permanently changed by the reaction.)

2-107 Pepsin is a protease that is secreted into the human stomach. Pepsin would

(A) have an optimum pH of 2. (The optimum conditions for enzyme activity will be similar to the environment in which it is found. The stomach is at body temperature and gastric juices have a pH of about 2. Proteases breakdown proteins.)

Section II Questions

2-108

(a) Animal cell as there is not cell walls. Plant cells have cell walls whereas animal cells do not.

(b)

mitochondrion	major site of ATP production
ribosome	**site of protein synthesis**
nuclear membrane	**controls what enters and leaves the nucleus**
Golgi body	packages material for secretion out of the cell

(c) Stains are taken up differentially by structures in cells. This improves contrast enabling structures to be seen more clearly. Stains can also be used to locate specific compounds in cells.

2-109

(a) Early microscopes were light microscopes. Light microscopes use glass lenses to focus light onto an object and then collect the transmitted light. Early microscopic images were limited in their magnification and resolution. Magnification refers to the size of the image and resolution refers to the ability to distinguish fine detail. Improvement in lenses, sectioning techniques and staining resulted in improved magnification and resolution, as a result, chloroplasts could be seen with light microscopes, but their fine detail could not.

In 1933 Ernst Ruska built the first electron microscope. Electron microscopes use beams of electrons instead of light. Electrons pass through preserved samples in transmission electron microscopes and are bounced off the surface of preserved samples in scanning electron microscopes. Electron microscopes have far greater magnification and resolution. They allow the fine detail of organelles to be observed. The fine structure of chloroplasts has been studied using electron microscopes. A major disadvantage in the use of electron microscopy is that only preserved material can be seen.

In 1934 Fritz Zernike invented the first phase contrast microscopes. These microscopes amplify small phase shifts in light passing through transparent material. They allow unstained living material to be seen in great detail. These microscopes have been used to study the movement of chloroplasts in cells, the three-dimensional structure of chloroplasts and the structure of grana.

More recently fluorescence microscopes have been used to study chloroplast development and the location of chlorophyll in chloroplast lamellae. Fluorescence microscopes use a very high intensity of light to illuminate the sample. This causes chlorophyll or fluorescent dyes in the sample to fluoresce.

Therefore, improvements in microscopy have enabled scientists to improve their understanding of how chloroplasts behave in vivo and increase knowledge of the detailed internal structure of chloroplasts.

(b) Information from secondary sources, such as books, research papers, and the internet, were used to gather information about chloroplasts and microscopes. Initially, key words were typed into a search engine. An article from Wikipedia was used as a starting point. The information given in the article was summarised under headings, which were then used to set up a table. The headings were placed along the top and other references along the side. Information was then gathered from a range of books, journals and websites.

(c) The initial key questions were used to help determine the relevance of the information found. Reliability was determined by whether the information agreed with other sources, a recent date of publication, and the recognised expertise of the author and the recognised reliability of the website (a well-known university.edu).

2-110

(a) A = phospholipid B = protein

(b) Both osmosis and diffusion involve the passive movement of substances from where they are in high concentration to where they are in low concentration. Diffusion refers to the movement of any gas, liquid or solute whereas osmosis refers to the movement of free water. Osmosis is the diffusion of water. Osmosis involves the movement of free water across a semi-permeable membrane whereas a membrane is not required for diffusion to occur.

2-111

Aim: to determine the concentration of saline solution that is isotonic to the sand worms' tissue.

Equipment list: 100 sand worms, scales, 5 beakers, 4 saline solutions -0.5, 1.0, 1.5 and 2.5 molar concentration

Method:

- Take 100 sand worms of similar age, size and health. Divide the worms into 4 groups of 25.
- Weigh each group and calculate the average weight per worm in each group.
- Place each group into a separate beaker. To the first beaker add 500 mL of 0.5 M saline solution, to the second beaker add 500 mL of 1.0 M saline solution, to the third beaker add 500 mL of 1.5 M saline solution and to the fourth beaker add 500 mL of 2.5 M saline solution
- Leave the worms in the solution for one hour and then reweigh. Calculate the average weight change for a worm in each solution.
- Repeat the experiment many times.

Risk assessment: wear gloves and watch for wave action when collecting the sand worms.

Independent variable: the different concentrations of the saline solution.

Dependent variable: average change in worm weight.

The saline solution in which no change in weight is observed will be isotonic to sand worm tissue.

Repeating the experiment ensures the reliability and validity of the data.

(When asked to design an experiment always include: aim; equipment list; data collection method; risk assessment; independent variable; dependent variable; predictions and repeating.)

2-112

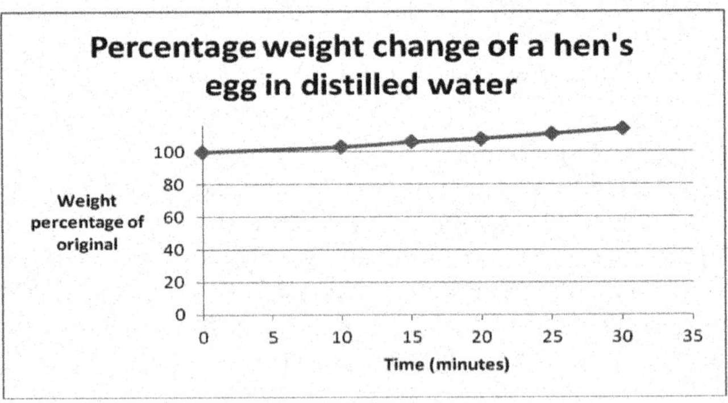

(Graph must:
- include a title
- include labelled axis including units
- be accurately plotted.)

(b) The egg gained weight at a steady rate as time in distilled water increased.

(c) The egg gained weight as water there was a net uptake of water into the cell. There was a higher solute concentration inside the cell than in the surrounding distilled water, as a result there was a net movement of free water across the semi permeable membrane of the egg into the cell.

2-113

(a) Diffusion

(b) Cube X

(c) Cube X had the highest surface area to volume ratio therefore the amount of acid moving in, compared to the volume of the block will be greatest.

2-114

(a) Chloroplast

(b) Inputs: water, carbon dioxide and light
Outputs: oxygen, glucose

(c) Any one of the following:
- temperature
- type/age/size of plant
- amount of water
- air flow

(d) 10 arbitrary units of light

(e) At Point Y carbon dioxide concentration and not light intensity is the limiting factor.

(f) No. The plant is respiring at the same time it is photosynthesising. Some of the oxygen produced in photosynthesis is used in respiration therefore the total rate of photosynthesis would be higher.

2-115

(a) An organic catalyst is a protein produced by a living organism. Its function is to speed up reactions. Amylase speeds up the breakdown of starch to maltose and water.

(b) To investigate the effect of temperature on amylase activity or to find the optimum temperature for amylase activity.

(c) The independent variable is the different temperatures tested and the dependent variable is the measurement of rate of maltose production.

(d) Any one of the following: concentration of amylase, concentration of starch, pH of the media, nutrients in the media.

(e) Validity would be improved by including an experimental control group. This group would be exactly the same except it would lack the amylase. This would show that the change in temperature alone caused the change in rate of maltose production.

(f) The optimum temperature for amylase activity is 40°C.

(g) No. Reliability refers to whether the results can be repeated. This experiment was performed once, therefore the results are not reliable.

Chapter 3 Organisation of Living Things
Content Review Questions – Suggested Answers

3-1 A large size protects an organism from predators.

3-2 As cells increase in size their SA:V decrease therefore their ability to obtain requirements and remove wastes decreases. A large organism comprised of many small cells each with a high SA:V ratio, reduces this problem. It also allows groups of cells to be specialised and differentiated leading to different cells functioning more efficiently.

3-3 A specialised cell is a cell that performs a particular function whereas a differentiated cell is a cell that performs a particular function and has a different appearance.

3-4 In multicellular organisms, a group of similar cells that work together to perform a particular function form a tissue. A defined structure composed of several different tissues grouped together to perform a particular function is called an organ and groups of organs working together to perform a general function are called a system.

3-5 Coordination. The cells within the multicellular organism must be able to coordinate their activities so that the whole organism functions efficiently.

3-6

	Cells within the following organisms		
	Unicellular	**Colonial**	**Multicellular**
Example	*Chlamydomonas*	*Volvox*	*Ulva* (sea lettuce)
Number of cells	One	Few to many	Many
Ability to undergo cellular respiration	Yes	Yes	Yes
Degree of specialisation and differentiation	-	Low	High
Ability to perform all functions in order to sustain life	Yes	Yes	No
Ability to survive alone	Yes	Yes	No

3-7 The internal environment of a unicellular organism refers to the conditions within its cell membrane whereas the internal environment of a multicellular organism refers to the fluid in which the cells bathe, e.g. tissue fluid, lymph and blood plasma. The external environment of a unicellular organism refers to the conditions outside its cell membrane whereas the external environment of a multicellular organism refers to the environment surrounding the whole organism.

3-8

Cell	Function	Observation	Explanation
Palisade cell	Photosynthesis	Many chloroplasts present	Chloroplasts are the site of photosynthesis.
Pancreatic cell	Secretion of insulin	Many Golgi bodies and mitochondria	Mitochondria provide energy for the production of insulin. Golgi bodies package the insulin for secretion.
Small intestinal	Absorption of products of digestion	Presence of microvilli	Microvilli provide a large SA for absorption of the products of digestion.

3-9

Flowering Plant System	Organs	Function
Shoot system	Leaves Stems Buds Flowers Fruit	Photosynthesis Gas exchange Transport Support Reproduction
Root system	Roots	Anchorage Gas exchange Uptake of water Uptake of mineral ions

3-10 The amount and types of nutrients required by an organism depends on:
- is it autotrophic or heterotrophic
- stage of growth of the organism
- level of activity of the organism
- reproductive state of the organism
- size of the organism
- environment that the organism lives in.

3-11 Effective gas exchange surfaces:
- have a large surface area
- are moist
- are thin
- are associated with a transport system that maintains the diffusion gradient
- are ventilated.

3-12 Roots are the site of water uptake in plants. Water is reactant in photosynthesis. Terrestrial plants are not supported by water. Roots anchor the plant and provide support for the stem. This increases the area exposed to sunlight and increases photosynthesis.

3-13 Roots are long and thin providing a large surface area for exchange. The epidermis also contains root hairs. These cells have long thin projections that further increase surface area. Roots are naked. This means they are not covered by a waterproof layer that would reduce exchange.

3-14 There is a higher solute concentration inside the root than in the surrounding soil. Water then moves by osmosis across the semi-permeable membrane of the root hair into the root.

3-15 Where the concentration of a particular mineral ion is greater in the soil than in the root, the ion is taken up by diffusion. Where the concentration of the ion is greater in the root than in the soil, it is taken up by active transport.

3-16 Root hairs allow gas exchange to occur in roots. Roots do not photosynthesise as they are underground. Roots require oxygen and produce carbon dioxide in aerobic respiration. Root hairs are naked and have a high SA:V allowing oxygen to pass into the root and carbon dioxide to pass out.

Lenticels allow gas exchange to occur in woody stems. They are small raised pores that do not have mechanisms for closing. Oxygen diffuses through the lenticel to the underlying parenchyma cells and carbon dioxide diffuses out.

3-17 Structural adaptations of leaves for photosynthesis:

Structural adaptation	**Explanation**
Shape	Leaves are usually broad, flat and thin in shape. This increases the surface area where light is absorbed and provides a large area in which stomata for gas exchange can be found. The leaf is thin allowing gases to be transported efficiently by diffusion.
Clear (transparent) cuticle and epidermis	Allows light to penetrate to the photosynthetic tissue. (Also protects the plant from excess heat gain and water loss. If the plant loses too much water the stomata, will close and less CO_2 will be taken up.)
Palisade layer	The palisade layer is near the surface increasing the plants ability to trap light. The palisade cells are rectangular and contain many chloroplasts. The chloroplasts contain chlorophyll, the pigment that traps light.
Stomata	Stomata allow CO_2 and O_2 to be exchanged between the leaf and the air. Usually more are found on the bottom of the leaf to reduce water loss. Evaporation of water from the leaf helps transport water to the photosynthetic cells.
Spongy mesophyll	Spongy mesophyll cells contain chloroplasts and can photosynthesise. The many air spaces between the cells allow CO_2 to diffuse rapidly to the palisade layer.
Vascular bundle - xylem - phloem	Water is transported to the photosynthetic cells via the xylem. Glucose is produced in photosynthesis. It is converted to sucrose and then transported out of the leaf to areas of storage and use by the phloem.

3-18 Stomata (plural) are usually found in the lower epidermis of leaves. One stoma (singular) is formed by two guard cells. The guard cells are banana shaped and the aperture gap between the guard cells can vary in size. The two guard cells are joined at the ends, their cell walls are thicker on the side next to the stoma and bands of radial microfibrils run around each cell wall.

3-19 Guard cells, unlike other cells in the lower epidermis, do have chloroplasts. Epidermal cells are transparent and let light through to the photosynthetic cells. The fact that guard cells have chloroplasts means that they can photosynthesise in light. Photosynthesis produces glucose which alters the osmotic potential of the guard cell compared to surrounding epidermal cells. As a result, water moves into the guard cell, opening the stoma and allowing gas exchange for photosynthesis by the mesophyll.

3-20 Stomata open and shut due to changes in the **concentration of solutes** in the guard cells. Photosynthesis produces glucose, which increases the solute concentration of guard cells. As a consequence, water diffuses into guard cells by **osmosis** making the guard cells **turgid**. This increases the bend and hence the stomatal aperture (opening). Other mechanisms may explain stomatal aperture. The plant, through production of hormones such as abscisic acid, can alter stomatal aperture. The result in all cases is a **change in solute concentration** of guard cell contents compared to surrounding epidermal cells. If the **concentration of solutes** in guard cells **drops** below that in the epidermal cells, water will move from the guard cells to the epidermal cells leaving the guard cells **flaccid** and stomatal aperture reduced. (Other factors that open stomata are plenty of water, light and low internal carbon dioxide concentration.)

3-21 Photosynthesis increases glucose concentration in photosynthetic cells such as the guard cells. Increased solute concentration causes water movement into guard cells which results in increased stomatal aperture.

3-22 Carbon dioxide, oxygen, water vapour and anything else in the air could pass through open stomata. Usually we are concerned with the first three.

3-23 The major factor that determines the rate of gas movement is concentration. Carbon dioxide diffuses into photosynthesising leaves. Oxygen diffuses out. The pattern is reversed at times when the leaf is not photosynthesising. Concentration gradients for these two gases will be determined by the rate of respiration compared to the rate of photosynthesis. Water always diffuses out as the intercellular spaces of leaves are saturated (humidity is 100%). The concentration of water in the air cannot be higher.

3-24 On a warm sunny day (plant photosynthesising) oxygen and water will diffuse out of the leaf while carbon dioxide will diffuse into the leaf. Movement is down each gas' concentration gradient.

3-25 Plants require carbon dioxide to photosynthesise. The carbon dioxide diffuses into the leaf through open stomata.

3-26 Plants tend to reduce stomatal aperture during the hottest part of the day in dry conditions. Stomatal aperture is also reduced at night. Both strategies will reduce water lost through transpiration.

3-27

Part	Structure and Function
Nose	The nasal cavity is lined with hairs that filter dust and other items from the air. It is also well supplied with blood vessels that warm the incoming air.
Mouth	An alternate pathway for air to the lungs.
Trachea	Trachea is a tubular stricture supported by cartilage rings (c-shaped) that prevent the trachea from collapsing as air is drawn in. Carries air from the back of the mouth to the two bronchi.
Bronchus	Similar structure to the trachea that carry air to the two lungs.
Bronchioles	Smaller tubes that branch and carry air to individual alveolar sacs.
Alveolar Ducts	Fine ducts leading from the bronchioles to individual alveoli.
Alveoli	Thin walled spherical structures that provide the gaseous exchange surfaces in the lungs.

3-28 Features of alveoli that enhance gas exchange:

- large surface area.
- thin. Only one cell thick providing only a small barrier to the movement of gases across the surface.
- alveoli are within the body therefore are able to be kept moist for efficient gas exchange.
- surrounded by capillaries the walls of which are only one cell thick providing a minimum barrier to movement of gases.
- blood within the capillaries is moving thereby maintaining a diffusion gradient.

3-29 Ventilation is the movement of air in and out of the lungs. Its purpose is to maintain a higher concentration of oxygen in the alveoli than in the nearby capillaries, so oxygen diffuse from the alveoli to the blood in the capillary network surrounding the alveoli. Ventilation also removes carbon dioxide from the lungs, which helps in the diffusion of carbon dioxide in the blood to the alveoli and out in exhaled air.

3-30 Air moves in and out of the lungs due to changes in the pressure inside the lungs. Pressure inside the lungs is decreased when the diaphragm contracts (moves down) and the intercostal muscles (between the ribs) also contract. This increases the volume of the chest cavity decreasing the pressure in the lungs below the external air pressure – air flows in. When the diaphragm and intercostal muscles relax pressure in the lungs is increased causing air to flow out.

3-31 Air entering the lungs has a higher oxygen concentration and a lower carbon dioxide concentration. Air leaving the lungs has a lowered oxygen concentration and a higher carbon dioxide concentration.

3-32

	Insect	Fish	Frog	Mammal
Diagram	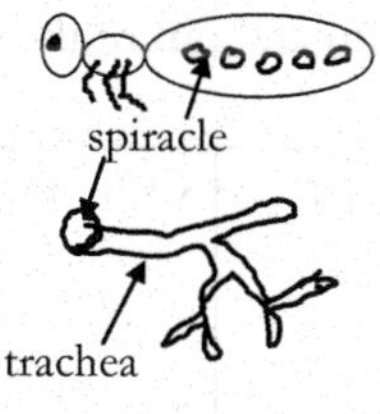 spiracle trachea	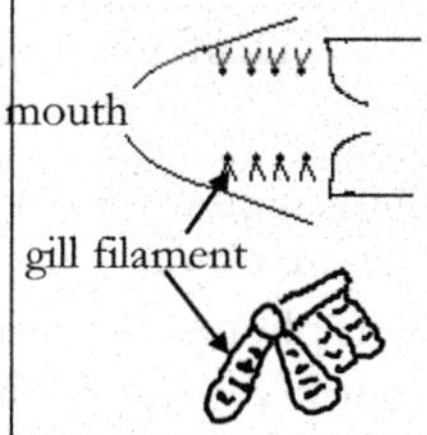 mouth gill filament	skin simple lungs	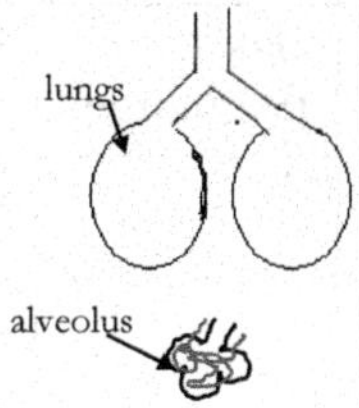 lungs alveolus
Description	A system of branching tubes (tracheae) extending into the insect's body from openings called spiracles.	Each gill consists of two rows of gill filaments. The gill filaments are covered with folds called lamellae.	The body-surface. Lungs: simple with few alveoli. No diaphragm	Sac–like organs. Air enters the mouth, to the trachea which divides and divides into bronchioles. The bronchioles end in alveoli.
Structure where exchange occurs	Tracheole ends	Lamellae	Body surface and alveoli	Alveoli
How the surface is kept moist	Surface of exchange is within the body. The spiracles sometimes have valves that maybe closed.	The gills are in water.	Must live in moist environments	Surface of exchange is within the body.
Ventilation	Muscle contractions of the body	Water flows in the mouth, over the gills and out the gill slits. Swimming or movement of the floor of the mouth aid flow.	Lung and mouth breathing	Inspiration: ribs rise, diaphragm moves down. Expiration: rib slower and diaphragm moves up.
Maintaining the diffusion gradient	Open circulatory system	Single circulatory system	Closed circulatory system (3 chambered heart)	Closed circulatory system (4 chambered heart)

3-33

		Plants	**Mammals**
Gas Exchange	**Role**	Transfer gases into and out of the organism	
	Gases	Photosynthesising tissue: CO_2 in; O_2 out Non-photosynthesising tissue: CO_2 out; O_2 in	CO_2 out; O_2 in
	Structures	Root hairs Lenticels Stomata	Lungs/alveoli
	Process	Diffusion	Ventilation Diffusion

3-34 Usually the molecules in food are too big to pass across the membranes of cells lining the digestive tract. They must be broken down or digested into molecules that are small enough to do this.

3-35 Effective digestive systems share the following characteristics:
- structures that enable efficient physical digestion
- the sequential production of enzymes for efficient chemical digestion
- a structure for storage of food while chemical digestion occurs
- a large surface area for absorption
- a transport system that maintains the diffusion gradient
- structures for the efficient removal of undigested wastes.

3-36 Physical digestion occurs when a starch biscuit is crushed or broken into small pieces of starch. These small pieces have not been chemically changed, as they are still molecules of starch. In chemical digestion, enzymes catalyse the chemical reaction where starch is broken down to form smaller maltose molecules. In summary, chemical digestion involves enzymes whereas physical digestion does not. Chemical digestion results in different molecules of smaller size whereas mechanical digestion does not.

3-37 Extracellular digestion involves the secretion of enzymes from the cell. Chemical digestion then occurs outside the cell. Intracellular digestion occurs within the cell. Large food molecules are taken into the cell by phagocytosis. They form food vacuoles within the cell. Enzymes are released into the food vacuoles resulting in the food being digested within the cell.

3-38 Physical digestion results in many small pieces of food. These small pieces have a higher combined surface area than the original piece. This means that there is more area exposed to the enzymes and therefore a faster reaction.

3-39

Structure	Major functions	Secretions	Mechanical digestion	Chemical digestion
Mouth teeth tongue	Ingestion Mechanical digestion Beginning of chemical digestion of carbohydrates	Saliva - amylase - lysozyme Mucus	Teeth and tongue	Carbohydrates Starch to maltose (amylase)
Oesophagus	Transport	Mucus	Some during peristalsis	
Stomach thick walls	Chemical digestion of proteins Mechanical digestion Storage	Gastric juice - HCl activates - pepsinogen to pepsin (protease) - lipase Mucus	3 muscle layers allow the stomach to contract in many directions	Proteins Some lipids
Small intestine	Final digestion of proteins, lipids and carbohydrates Absorption - glucose - amino acids - fatty acids and glycerol - water - mineral ions - vitamins	Pancreatic and intestinal juices - amylase - maltase - protease - lipase - alkaline Bile Mucus	Bile emulsifies fats Walls during peristalsis	Proteins Carbohydrates Lipids
Large intestine	Absorption of salts, water and some vitamins Formation of faeces	Mucus		
Rectum and anus	Elimination	Mucus		

3-40 Bile emulsifies fats. This means it breaks fatty layers into small droplets with high surface area to volume ratio. These droplets are not chemically altered so the action of bile is an example of physical digestion.

3-41 No. The optimum pH for salivary amylase is 7.0–7.5 whereas the pH of gastric juice is about 2.

3-42 The lining of the small intestine is folded into structures called villi. Each villus is made of cells, the outer surface of which is folded into even smaller projections called microvilli. The presence of villi and microvilli increase the surface area of the intestine lining available for absorption.

3-43

Food group	Enzyme involved	Product of digestion	Absorbed across the wall of the small intestine into…
Carbohydrate	Amylase	Glucose	Blood
Protein	Protease	Amino acids	Blood
Lipids	Lipase	Fatty acids and glycerol	Lymph

3-44 Koalas mainly eat eucalypt leaves that contain large amounts of difficult to digest cellulose. Their caecum is large as it contains many symbiotic, cellulose-digesting bacteria. The bacteria digest the cellulose producing glucose for themselves and for the koala.

3-45

	Herbivore	Carnivore	Nectar feeder
Teeth	Flat incisors Small or absent canines Broad, rigged molars	Pointed incisors Large canines Jagged molars	Few small teeth
Length of digestive tract	Long	Short	Short
Stomach	May be large - foregut fermenters (sheep, kangaroos)	Medium	Two chambers where nectar and pollen are stored. Does not secrete proteases
Caecum	May be large - hindgut fermenters (horses, rabbits, possums, koalas)	Small or absent	None

3-46 Vascular plants obtain:

	Source	Structure involved in uptake
CO_2	Cellular respiration Air	 Stomata in leaves Lenticels in stems
O_2	Photosynthesis Air Air in soil spaces	 Stomata in leaves Lenticels in stems Root hairs
H_2O	Cellular respiration Soil water	 Root hairs
Mineral ions	Soil	Root hairs
Organic molecules	Reactions within the plant including photosynthesis and protein synthesis	Chloroplast Ribosomes Endoplasmic reticulum

3-47 Mammals obtain:

	Source	Structure involved in uptake
CO_2	Cellular respiration	Lungs
O_2	Air	Lungs
H_2O	Food and drinking	Digestive system
Mineral ions	Food	Digestive system
Organic molecules	Food	Digestive system

3-48 Large multicellular organisms have a low surface area to volume ratio. Internal cells are too far away from exchange surfaces to rely on diffusion. The rate of supply of requirements and removal of wastes to cells by diffusion would be too slow to sustain life.

3-49 Effective transport systems have
- tubes of various sizes allowing for rapid transport through large vessels and a site of exchange at small vessels with high surface area to volume ratio.
- a pump to push the fluid around the body
- separation of oxygenated and deoxygenated transport fluid
- extensive network with vessels for exchange close to all cells
- a fluid with high carrying capacity
- control mechanisms that allow fluid to be directed to areas of high metabolism.

3-50

	Xylem	Phloem
Diagram	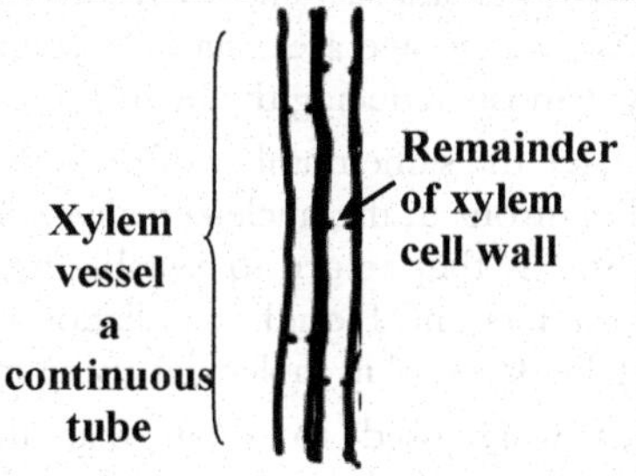 Xylem vessel a continuous tube; Remainder of xylem cell wall	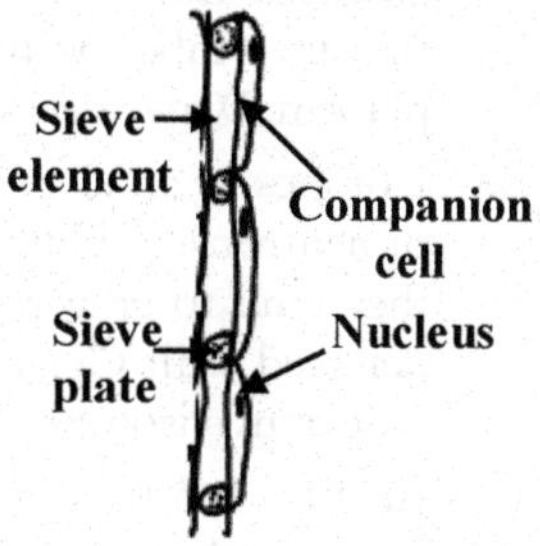 Sieve element; Companion cell; Sieve plate; Nucleus
Cell type/ vessel structure	Tracheid, xylem vessel	Sieve element, companion cell
Presence of cytoplasm	Absent	present
Presence of a nucleus	Absent	Not present in sieve elements. Present in companion cell.
Thickness of wall	Very thick	Thin
Alive/dead	Dead	Alive
Name of transport process	Transpiration stream	Translocation
Substances transported	Water Mineral ions up	Sucrose Most hormones Mineral ions down

Direction	One way Roots to leaves	Two-way
Energy source	Sun	Cellular respiration

3-51 Pathway from soil to leaf: soil solution, root hairs, root cortex (either around the cells or through the cells), endodermis, xylem in the root, xylem in the stem, xylem in the leaf, leaf intercellular spaces and finally lost to the atmosphere through open stomata.

3-52 Transpiration is evaporative water loss from the leaf surface. The sun provides the energy for the evaporation of water. As the water evaporates, it draws more water from the leaf vascular bundles by cohesion. Cohesion is the force of attraction between water molecules. This is transferred to water molecules in the xylem vessels resulting in the molecules being pulled up the stem. Water molecules are not only cohesive. They are also adhesive to the walls of the xylem vessels. This means that they are attracted to the walls of the small xylem vessels and this helps move the water up the plant.

3-53 The transpiration stream is the flow of water through the plant. Transpiration is the evaporative water loss from leaf surfaces. Most of this occurs out of open stomata. The two are related in that if there is no transpiration stream there will be no transpiration.

3-54 Sucrose produced in the leaves is actively transported into the phloem sieve elements. As the concentration of sucrose increases, water moves into the sieve elements from the phloem by osmosis. This increases the hydrostatic pressure in the sieve tube. At the root or storage area, sucrose is actively pumped out of the phloem. Water flows by osmosis reducing the hydrostatic pressure.

3-55 Isotopes of an element have the same number of protons but differ in the number of neutrons. Additional neutrons in the nucleus makes some isotopes unstable and they tend to emit radiation as they return to a stable form. This radiation can be detected with Geiger Counters and Liquid Scintillation Counters or heavier non-radioactive isotopes may be detected in molecules using mass spectrometry.

In the 1930s, isotopes were used to show that the oxygen produced in photosynthesis came from water and not from carbon dioxide. Isotopes can also be used to investigate each step in a biochemical pathway. Molecules containing heavier isotopes are introduced to cells and at short time intervals after the introduction, small samples are taken and the compound in which the isotope is found is isolated and identified. Using this method Melvin Calvin was able to work out the steps in the Calvin cycle.

Using radioactive carbon dioxide and aphids it has been shown that glucose produced in photosynthesis is converted into sucrose and transported throughout plants in the phloem.

3-56

	Arteries	Veins	Capillaries	Lymph vessels
Major function	To deliver requiremen ts to tissues	To remove wastes from tissues	Site of exchange between the blood and cells/external environment	Return excess tissue fluid to the blood (Have a major role in the defence against disease)
Thickness of wall	Very thick	Medium	Very thin (one cell thick)	Medium
Elasticity	Very	Medium	Low	Medium
Continuous/b lind ended	Continuou s	Continuous	Continuous	Blind
Presence of valves	No (except for at the beginning of the arteries leaving the heart)	Yes	No	Yes
Direction of flow	Away from the heart	To the heart	Connect arteries to veins	To the heart
Blood/ fluid pressure	High	Low	Low	Low
How the fluid moves	By contractio ns of the heart	By compression s of veins during body movements	By contractions of the heart	By compressions of lymph vessels during body movements
Oxygenated/ deoxygenated	Usually oxygenated except for pulmonary artery	Usually deoxygenate d except for pulmonary vein		

3-57 No, not all arteries carry oxygenated blood. Arteries carry blood away from the heart. The pulmonary artery carries blood away from the right-hand side of the heart to the lungs. It carries deoxygenated blood. Blood in arteries leaving the left-hand side of the heart is oxygenated.

3-58 Starting at the left ventricle:

Left ventricle, aorta, arteries to tissues of body (except the lungs), arterioles, capillary beds in tissues to allow exchange, venules, veins carrying blood towards heart, vena cava, right atrium of heart, right ventricle, pulmonary artery to lungs, capillaries surrounding alveoli, pulmonary veins carrying blood back to the left atrium, and to the left ventricle- completing the circuit.

3-59

Major blood component	Function
Plasma. The fluid part of the blood containing: water, glucose, amino acids, fatty acids, vitamins, plasma proteins, nitrogenous waste, carbon dioxide, oxygen (small amount), ions and hormones.	Transport of requirements to cells and wastes away from cells.
Cellular. - erythrocytes (red blood cells) - leucocytes (white blood cells) - platelets.	 - transport of oxygen - defence against disease - formation of clots.

3-60

		Method of transport
	Glucose	As part of blood plasma
	Amino acids	As part of blood plasma
	Fatty acids and glycerol	As triglycerides/chylomicrons in the lymph and then blood plasma
	Oxygen	Most is transported in red blood cells, bound to haemoglobin (oxyhaemoglobin)
	Carbon dioxide	Mainly as bicarbonate ions in the plasma (Note: the carbon dioxide enters the red blood cell where the enzyme carbonic anhydrase catalyses the formation of bicarbonate ions.)
	Nitrogenous waste	Mainly as urea as part of blood plasma
	Ions	As part of blood plasma
	Hormones **- water soluble** **- steroids**	 As part of blood plasma Bound to carrier proteins in the plasma

3-61 If oxygen did not bind reversibly with respiratory pigments, it would not be released and made available to the cells.

3-62 The amount of oxygen bound to haemoglobin depends on
- oxygen concentration in the solution (the higher the concentration the greater the saturation)
- carbon dioxide concentration in the solution (the higher the concentration the lower the saturation)
- pH of the solution (the lower the pH the lower the saturation)
- temperature (the higher the temperature the lower the saturation).

3-63

Concentration of	Capillary bed	
	Blood entering	**Blood leaving**
Oxygen	High	Low
Carbon dioxide	Low	High
Glucose	High	Low
Urea	Low	High

3-64 In open circulatory systems, the blood does not spend all the time in blood vessels. There is no difference between tissue fluid and blood. This is not true for a closed circulatory system. The blood and the tissue fluid have a different composition and the blood remains within the blood vessels. Closed circulatory systems are more efficient as the blood can be directed to all cells.

3-65 In the fish single circulatory system, the blood leaves the heart and moves to the gill capillaries where oxygen is taken up and carbon dioxide removed from the blood. The blood then moves into larger vessels and transported to capillary beds for exchange with cells. The heart must pump the blood through at least two capillary beds. In mammals the heart pumps blood to the lung capillaries and then back to the heart. The left-hand side of the heart then pumps blood to the body cells without passing through the lung capillaries. When blood passes through capillaries, blood pressure drops therefore in a single circulatory system oxygen-rich blood leaving the gas exchange surface will be transported less efficiently than in a double circulatory system.

3-66

		Plants	**Mammals**
Transport systems	**Role**	Bring requirements to cells and to take waste to excretory organs	
	Number of separate systems	2 separate systems; xylem and phloem	One closed system
	Structures	Vascular bundles -phloem -xylem	Heart Arteries Veins Capillaries Lymph vessels
	Substances transported	Sucrose Mineral ions Plant growth regulators	Products of digestion O_2 CO_2 Mineral ions Hormones Nitrogenous waste Antibodies
	Energy source	Xylem – sun Phloem – cellular respiration	Cellular respiration – contraction of heart muscles
	Direction	Xylem – up from the roots to the leaves Phloem – up and down	One direction

Section I Questions

3-67 Which of the following is a list from most complex to simplest?

(D) multicellular, colonial, unicellular. (Unicellular organisms consist of one cell; colonial organisms are a loose collection of cells whereas multicellular organisms contain many cells that maybe specialised and differentiated.

3-68 Specialised cells

(B) perform a specific function and may look different from other cells.

(Specialised cells perform a specific function. They may or may not be differentiated from other cells.)

3-69 It is true to say

(D) both cells are specialised and differentiated. (Cell A is specialised to perform photosynthesis and Cell B is specialised for absorption. Both cells look different therefore both cells are specialised and differentiated.)

3-70 Which of the following is not a characteristic of an efficient surface for gas exchange?

(D) thick epidermal cell membranes. (Exchange surfaces are thin, moist, have a high surface area and maintain a high diffusion gradient.)

3-71 A possible reason for this is

(D) the root cells are unable to get enough oxygen for aerobic respiration and the cells die. (There is less oxygen in water and the water in the soil displaces the oxygen in the soil spaces. Roots must respire in order to live and provide the energy for active uptake. Roots are underground, so they are not able to photosynthesise.)

3-72 Adaptations of leaves to increase photosynthesis include

(A) a large surface area. (A thick waterproof cuticle, stomata in pits and hairs on the surface are adaptations to reduce water loss.)

3-73 Within leaves

(A) the palisade layer is the major site of photosynthesis. (The spongy mesophyll layer contains many air spaces and some photosynthesis occurs here. Usually epidermal cells do not contain chloroplasts. Guard cells do contain chloroplasts.)

3-74 As a result

(A) the size of the stoma will increase. (Water moving into the guard cells increases the size of the pore or stoma.)

3-75 Gas exchange occurs across

(A) alveoli in mammals. (Gas exchange occurs across lamellae in fish, tracheoles in insects and alveoli in whales.)

3-76 In the lungs

(A) air sacs called alveoli increase the surface area for gas exchange. (Air is drawn in when the diaphragm contracts, oxygen diffuses into the blood and 100% of the inhaled oxygen does not pass into the blood.)

3-77 Gills have

(A) a counter current system to increase gas exchange. (Gills have a high SA as water has less oxygen available to be absorbed compared to air. A protective barrier would reduce oxygen uptake. Oxygen is absorbed from the water.)

3-78A disadvantage of obtaining oxygen from air is

(B) water loss across the respiratory surface may lead to dehydration. (There is more oxygen in air than in water therefore less surface area is required.)

3-79 Heterotrophs require digestive systems because

(B) the molecules in the food they eat are too big to pass into the body. (Heterotrophs are able to build organic molecules, but they must consume the basic building blocks of these molecules, e.g. glucose, amino acids and fatty acids.)

3-80 In humans, carbohydrate digestion begins in the

(A) mouth. (Fact.)

3-81 Koalas have a large caecum that is used to

(D) digest cellulose. (The caecum contains cellulose-digesting bacteria. The largest part of a koala's diet is gum leaves containing a high percentage of cellulose.)

3-82 In humans, bile

(B) is involved in the mechanical breakdown of lipids. (Bile is made by the liver, stored in the gall bladder and it is not an enzyme. It is involved in the mechanical digestion of lipids. Bile is an emulsifier.)

3-83 Transpiration will be highest for a well-watered pot plant when the

(A) air is moving, and the plant is in bright sunlight. (Moving air helps maintain the diffusion gradient and bright sunlight implies high temperature therefore the greater the rate of evaporation of water from the leaf surface.)

3-84 With respect to transport within vascular plants, one of the following is true.

(C) It requires the sun's energy for movement of water. (Occurs in two directions in the phloem, which are living cells. Carbohydrate is transported as sucrose not glucose.)

3-85 The blood vessel is

(C) an artery. (Arteries have a thick walls and small lumen.)

3-86 Blood returning to the heart via the pulmonary vein

(C) enters the left atrium, flows to the left ventricle and then to the rest of the body via the aorta. (The pulmonary vein carries blood from the lungs to the left side of the heart. Blood leaves the left side of the heart and moves to the body via the aorta.)

3-87 In the lungs

(A) oxygen diffuses from the blood in the capillary into the alveolus. (In the lung oxygen diffuses from the alveoli into the blood in the capillary. Carbon dioxide diffuses from the blood in the capillary into the alveolus. There is no blood in the alveolus.)

3-88 pH decreases when CO_2 dissolves in water therefore

(B) blood entering a capillary bed will have a higher pH than blood leaving a capillary bed. (Blood entering a capillary bed will contain less carbon dioxide and therefore higher pH than blood leaving a capillary bed. The pulmonary vein carries blood with less dissolved CO_2 and therefore higher pH.)

Section II Questions

3-89

(a) Any one of
- nucleus
- mitochondrion
- endoplasmic reticulum

(b) *Chlamydomonas*:
Carbon dioxide: obtained by diffusion across the cell membrane.
Water: obtained by osmosis across the cell membrane.
Light: an eyespot detects light and the flagella move the organism into the light.
Palisade cell:
Carbon dioxide: obtained by diffusion through the stomata, across the spongy mesophyll and across the cell membrane.
Water: from the soil via the root hairs, root, xylem vessels in the stem and leaf and osmosis across the cell membrane.
Light: the position of the leaf allows sunlight to pass across the epidermis into the palisade layer.

(c) No. *Chlamydomonas* can live independently whereas the palisade cell cannot.

(d) Any one of palisade mesophyll, spongy mesophyll, epidermis, xylem, phloem.

3-90

(a) Five major features of the lung are:
- very large surface area for exchange.
- moist surface to allow oxygen to dissolve which allows diffusion.
- thin walls (alveoli walls are one cell thick) to reduce distance from exchange surface to transport vessels
- a concentration gradient from alveoli to blood, which is maintained.
- a rich supply of capillaries near the exchange surface.

(b) Breathing can become frequent – lungs ventilated more often.
Breaths can become deeper – changing a greater volume of the air in the lungs per breath.
Both of the above help to maintain the concentration gradient in the lungs so that oxygen can get to actively respiring tissues more efficiently.

(c) Haemoglobin in red blood cells, allows more efficient oxygen transport in the blood. At the lung capillaries, oxygen and haemoglobin combine to form oxyhaemoglobin.

(d) Carbon monoxide combines with haemoglobin even more efficiently than does oxygen and it is hard to reverse the process. Carbon monoxide effectively prevents the transport of oxygen, so cells die due to a lack of oxygen.

3-91

(a) The digestive system brings about the **breakdown** of ingested food into small enough molecules/particles that can be **absorbed** through the intestine walls into blood vessels to be transported to where they are needed around the body.

(b) Breakdown in the mouth by teeth is mechanical digestion. Most of the breakdown in the small intestine is due digestive enzymes and is therefore chemical digestion.

(c) The role of the small intestine is to absorb the products of digestion such as: glucose and amino acids. To speed up the process the wall of the intestine is folded, and the folds are covered in projections called villi, which in turn are covered with microvilli. This increases the surface over which absorption can occur.

(d) The major role of the large intestine is in absorption of water and minerals. The large intestine is thin walled and well supplied with blood vessels.

3-92

(a) Animal S

(b) Animal S does not have canine teeth. Canine teeth are used for tearing meat therefore are normally present in carnivores not herbivores. Animal S has a large caecum. Cellulose is difficult to digest and is a large component of an herbivores diet. A large caecum increases the animal's ability to digest cellulose as cellulose digesting bacteria can be found in the caecum. Finally, as cellulose is hard to digest a longer intestine is required than for a carnivore.

3-93

(a) The use of isotopes to track the movement of atoms through plants has been of major importance in the study of plant transport systems and photosynthesis. Before the use of isotopes this was very difficult as atoms cannot be seen even with microscopes and biochemical pathways are very complicated involving many intermediate steps.

Isotopes of an element have the same number of protons but differ in the number of neutrons. Additional neutrons in the nucleus makes some isotopes unstable and they tend to emit radiation as they return to a stable form. This radiation can be detected with Geiger Counters and Liquid Scintillation Counters or heavier non-radioactive isotopes may be detected in molecules using mass spectrometry.

In the 1930s, isotopes were used to show that the oxygen produced in photosynthesis came from water and not from carbon dioxide. Isotopes can also be used to investigate each step in a biochemical pathway. Molecules containing heavier isotopes are introduced to cells and at short time intervals after the introduction, small samples are taken and the compound in which the isotope is found is isolated and identified. Using this method Melvin Calvin was able to work out the steps in the Calvin cycle.

Using radioactive carbon dioxide and aphids it has been shown that glucose produced in photosynthesis is converted into sucrose and transported throughout plants in the phloem.

Therefore, it can be seen that modern techniques, have allowed biologists to study the source of oxygen produced in photosynthesis and transport in the phloem.

(b) Information from secondary sources such as books, research papers, and the internet were used to gather information about oxygen, photosynthesis, and transport in the phloem. Initially, key words were typed into a search engine. An article from Wikipedia was used as a starting point. The information given in the article was summarised under headings, which were then used to set up a table. The headings were placed along the top and other references along the side. Information was then gathered from a range of books, journals and websites.

(c) The initial key questions were used to help determine the relevance of the information found. Reliability was determined by whether the information agreed with other sources, a recent date of publication, and the recognised expertise of the author and the recognised reliability of the website (a well-known university.edu).

3-94

(a) Arteries have thick muscular walls that contract to maintain high blood pressure. Veins have a thin layer of muscle and rely on the presence one-way valves to maintain blood flow back towards the heart.

(b) Capillary walls are one cell thick compared to the multi-layered arties and veins. Capillaries are thin-walled to allow movement of materials in or out depending on location. Sometimes materials will diffuse into capillaries – oxygen at the lungs. Sometimes things will diffuse out – oxygen at active muscle tissue.

(c) The pulmonary arteries carry deoxygenated blood to the lungs. Other arteries carry oxygenated blood to the rest of the body. The pulmonary veins carry oxygenated blood to the left side of the heart. Other veins carry deoxygenated blood back to the right side of the heart.

(d) If there is a 'hole in heart' some blood can pass from the right side of the heart to the left side without passing through the lungs. This means that blood leaving the left ventricle for the body will not be fully oxygenated. This affects the individual's capacity for sustained exercise.

(e) The normal red blood cell (a bi-concave disc) has a high SA/V ratio so can load and unload oxygen quickly. The sickle celled shaped red blood cells have a lower oxygen carrying capacity and tend to clump together blocking blood vessels.

Chapter 4: Biological Diversity
Content Review Questions – Suggested Answers

4.1 Abiotic factors are the non-living factors that impact on organisms. For example: water quality, light, temperature and soil conditions.
Biotic factors are the living factors that impact on organisms, for example; the presence of predators, prey organisms, diseases or competitors.

4.2 Comparison of the abiotic factors present in marine, freshwater and terrestrial environments.

Abiotic factor	**Environment**		
	Marine	**Freshwater**	**Terrestrial**
Availability of water	High but access depends on salinity/osmosis	High	Low
Availability of light	Intensity decreases with depth and different wavelengths vary in penetration. Red light has the least penetration.		High
Availability gases (oxygen/ Carbon dioxide)	Low as dissolved in water. More oxygen dissolves at low temperature and decreased depth		High
Availability of ions	High especially Na^+ and Cl^-	Low	Available in the soil. Type and amount vary according to type of soil.
Temperature	Little variation Rock pools – large variation	Depends on the size of the body of water. Large body – constant Small body –increased variation	Huge range. Depends on time of day, seasons, shelter, wind
Viscosity	High		Low
Buoyancy	High		Low
Pressure	High and increases with depth		Varies only slightly with altitude
Turbulence	Depends on currents and tides	Depends on how fast the water is flowing, i.e. some rivers flow quickly	Depends on wind, weather and exposure
Substrate	Water with sandy/rock bottom	Water with sandy/muddy/rock bottom	Soil, rock, sand

4-3 The abiotic factors most important in determining distribution and abundance of aquatic organisms are:
- size of the body of water. This affects variation in temperature and dissolved ions
- depth of water. This affects the quality and quantity of light penetration, amount of dissolved gases and pressure
- strength of tides, currents and waves.

4-4 The abiotic factors most important in determining distribution and abundance of terrestrial organisms are:
- availability of water
- temperature variation both daily and seasonal
- extremes of temperature
- exposure to wind
- support.

4-5 Biotic factors determining distribution and abundance of organisms are:
- availability of food
- number of predators
- presence of pathogens
- number of mates
- number of competitors.

4-6 Native forest – biotic & abiotic factors

Abiotic factors	Biotic factors
• air temperature	• tree species present
• rainfall	• age & height of trees
• soils composition	• undergrowth species
• soil water holding qualities	• 1st order consumers
• soil nutrients	• predators
• prevailing winds	• competitors
• depth of litter layer	• decomposers & detritivores present

4-7 Light penetrates little more than 30 metres into the ocean and it is diminished considerably from the 10-metre mark. Most photosynthetic life is where most of the light is – in the top few metres of the oceans and bodies of water.

4.8 The individuals that are most likely to survive in a particular environment are those with the best set of adaptations for that environment.

4.9 Only individuals that have the best set of adaptations to a particular environment survive therefore the abiotic and biotic environmental factors select those individuals that survive thereby providing a selection pressure.

4-10

Scientific name	*Rhinella marina*
General characteristics	Toad – amphibian Ground dwelling Grow to approximately 15 cm in length Poison glands on back
Diet	Mainly insects but also fish and reptiles
Where did the cane toad come from?	South and central America Introduced to Australia from Hawaii
When were they introduced?	1955
Why were they introduced and was this successful?	As a biological control against cane beetles. It was not successful as the toads could not jump very high, so the cane beetles moved to the top of the cane and were not eaten.
What features of the cane toad allowed the population to increase rapidly?	- highly toxic skin - lack of a predator/disease - rapid reproduction - one female can produce up to 30 000 eggs - well adapted to a wide range of environments
Environmental impact	- death of native predators - competition for shelter and food with native species

4.11

Scientific name	*Opuntia* species
General characteristics	Cactus Reproduce asexually –pads Reproduce sexually –flowers Fruit is eaten by birds
Where did the prickly pear come from?	Americas Introduced to Australia from Brazil
When were they introduced?	1788
Why were they introduced?	As a garden plant For fencing For the cochineal industry
What features of the prickly pear allowed the population to increase widely?	- no species present in Australia limiting its growth - well adapted to the Australian climate - can reproduce asexually from pads (pieces of plant tissue) - reproduce sexually with flowers - seeds are easily spread by birds -seeds are viable for years
When and how was the prickly pear population controlled?	In 1926 the Australian Government introduced a South American cactus moth (*Cactoblastis cactorum*) into Australia. The moths lay eggs in the prickly pear flesh. When the eggs hatch, the prickly pear is used as a food source. The prickly pear population was controlled in 6 years.

4-12 Adaptation types

Structural adaptations	Parts of a body, e.g. a wing for flying or fins on fish for swimming.
Physiological adaptations	Changes in function of parts of the body depending responses to change, e.g. heart rate & breathing rate vary depending on needs of the body.
Behavioural adaptations	Activity of whole organism to enhance its survival, e.g. lizard finding a warm spot in the sun to gain heat to raise body temperature to allow efficient metabolism.

4-13 Behaviours of Australian animals use to survive harsh desert conditions include: nocturnal habits (active at night), reduced activity during day living in a burrow, hibernating during hottest/driest part of the year, drinking urine and spreading out – presenting largest surface area possible from which to lose heat.

4-14 Physiological adaptations that Australian animals use to survive harsh desert conditions include: lowering of hair to reduce insulation, increased blood flow to body surface to increase heat loss, decreased metabolic rate to reduce heat generation, raised body temperature to decrease heat flow from environment, greater use of metabolic water, greater reabsorption of water by the kidneys and increased sweating/panting (if not limited by water loss).

4-15 Cold climate plants in winter and desert plants in the dry season both suffer from the same problem – lack of suitable conditions for photosynthesis. In extreme cold weather, water may not be available and light conditions are poor, so it is a better strategy to lie dormant in winter until better light and water conditions return with spring. In very dry conditions, leaf loss means less surface area from which water may be lost – again the better strategy is to remain dormant until conditions are better.

4-16 Strategies mammals use to gain heat or increase heat loss include:

Heat gain strategies	Heat loss strategies
• Increase metabolic rate • Move to warmer area, e.g. lie in the sun • Drink warm liquids	• Decrease metabolic rate • Increase sweating/panting • Move to cooler areas • Get wet • Lower hairs/remove clothes – decreasing insulation

4-17 Behaviours that poikilothermic animals exhibit to regulate body temperature include:

- moving to or from warmer/cooler areas
- sheltering from particularly cold or hot conditions
- becoming more active in cool conditions – some heat is generated
- reduction of activity in very hot conditions – to reduce any heat generation.

4-18 Land temperatures change quite dramatically whereas water temperatures change relatively slowly. As discussed enzymes function better within narrow temperature ranges. Therefore, homeothermy – the ability to maintain constant body temperature – is more useful on land than in water.

4-19 Sweating results in water loss, as well as heat loss. For desert animals, water may be in very short supply and they cannot afford to lose it even if it is to cool their bodies. Most desert animal exhibit behaviour that reduces heat gain. For example, they often live in burrows and exhibit nocturnal behaviour. Some animals allow body temperature to rise during the day above what might be the optimum and then during the night, heat is lost as the external temperature cools. Body temperature then drops lower than the optimum. This adaptation allows temperature regulation achievement with minimal water loss.

4-20 The major problem flowering plants have in hot climates is minimising water loss while still allowing uptake of carbon dioxide. Carbon dioxide enters through open stomata. Water is lost, especially in hot conditions, through open stomata. It is a matter of balance obtaining carbon dioxide from the atmosphere and retaining water – both are necessary for photosynthesis and thus for growth.

4-21 Adaptations that flowering plants have to cope with hot dry climates can include:

- narrow leaf shape
- leaves that roll in particularly dry conditions
- leaf surface shiny to reflect light and heat
- loss of leaves in the dry season
- thicker waxy cuticle than found in plants from more temperate areas
- hairs on leaves
- reduced numbers of stomata
- stomata in pits
- inverted stomatal rhythm – stomata open at night.

4-22 The features shown by populations are the result of natural selection. Natural selection acts over a long period of time. Those individuals that have characteristics that give them an advantage in a particular environment survive, reproduce and pass on the genetic information coding for these features to the next generation. Over time conditions change and different features maybe selected for or against. Some features may no longer offer an advantage but if they do not disadvantage the organism they may continue to be passed on even though they are not an adaptation to that particular environment. This makes it difficult to decide if a feature shown by an organism is an adaptation to the present environment or to a past environment.

4-23 Biological evolution is the cumulative change in characteristics of populations over successive generations. To evolve means to change. Biological evolution has occurred if there is change in the characteristics of a population over a number of generations. Biological evolution is the result of change. The process that brings the change about is 'Natural Selection'.

4-24 Biological evolution is a scientific theory. There is no such thing as a scientific fact! Some theories are better than others. Many biologists believe there is much evidence to support the evolution theory.

4-25 'Biological Evolution' occurs as the result of 'Natural Selection'. 'Evolution' refers to the changes in species over time. 'Natural Selection' is the process that results in the change.

4-26 For natural selection to occur there must be phenotypic variation (based on genetic variation) in a population. If all individuals were phenotypically identical, selection for better-suited phenotypes could not occur.

4-27 Natural selection occurs at the phenotypic level. The individuals with phenotypes that suit them best to the environment will tend to be selected. Less well-suited individuals tend not to survive and reproduce. As phenotype is largely determined by genotype, alleles that lead to better-suited phenotypes will become more common. The selection, however, is at the phenotypic level.

4-28 The genetic fitness of an individual depends on the environment (selecting agents) in which it lives. Polar bears are well adapted to life in the Arctic Circle. A polar bear placed in the Simpson Desert would not survive long!

4-29 Individuals that are well suited to their environment are more likely to reproduce. Poorly suited individuals are less likely to survive to reproduce. If they do reproduce, on average they would produce fewer offspring.

4-30 To pass genetic information on to the next generation an individual has to survive long enough to reproduce. One thing of which you can be certain is that your ancestors reproduced. You can thank them for your genetic inheritance!

4-31 Summary of Evolution by Natural Selection

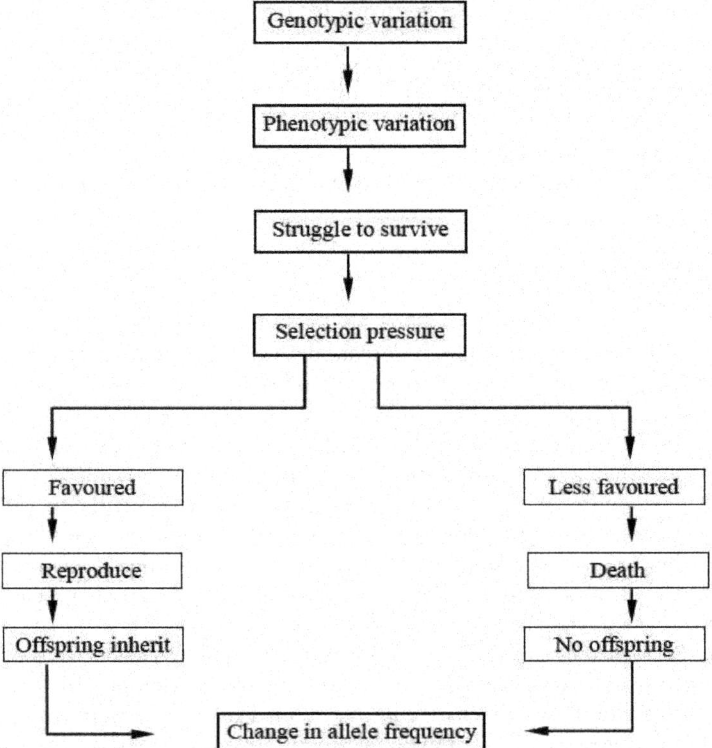

Within a species there is considerable **variation** in physical characteristics that are genetically determined. Some individuals are more likely to survive than others.

In nature, **large numbers** of offspring are produced but not all survive. There is a struggle for survival due to competition for limited resources. Only some individuals survive to reproduce.

Individuals with **certain characteristics** are more likely to survive to reproduce in a particular environment (selection pressure) than other individuals. The individuals that **survive** pass on their characteristics to the next generation. Those individuals with different characteristics are less likely to survive to reproduce. They are less likely to pass on their characteristics to the next generation.

Over time, there is an increase in the gene pool in the number of the alleles that determine the favoured characteristics.

4-32 Characteristics are coded for by an organism's genes. The gene pool is literally the pool of genes for a population. A population is a group of individuals of one species living in one area. Within a population there is genetic variation. That is, there are alternative forms of genes present – alleles. Calculation of frequency of particular alleles in the population's gene pool can occur. Over a number of generations, alleles that give individuals an advantage will increase in frequency in the gene pool.

4-33 There was original variation in colour in peppered moth populations. Some were dark, and others were pale. In polluted areas, tree trunks are dark. Dark moths are camouflaged and fewer are eaten by birds. These dark moths survive, reproduce and their offspring inherit the dark colouration. Pale moths are more obvious to birds, more are eaten and fewer survive to reproduce. Over time, the population has become predominantly dark. In unpolluted areas, tree trunks are pale. Dark moths are less camouflaged, are easily identified by birds and are eaten. Less survive to reproduce. Pale moths are more camouflaged. They are eaten less by birds therefore they survive, reproduce and their offspring inherit their pale colouration. Over time, the population has become paler.

4-34 The term 'species' is a human concept used to classify organisms. Classification is often subjective as organisms are hard to pigeonhole. Therefore, no definition is perfect!

A species is a group of organisms that are structurally and biochemically similar. Members share more characteristics in common than with any other individuals. Members of a species interbreed to produce vigorous fertile offspring. Members of a species share the same gene pool or are connected by gene flow to other populations' gene pools.

The species definition is a matter of debate and there is no absolute definition.

4-35 The test used most is to determine whether individual organisms mate to produce fertile offspring. This test is not always possible!

4-36 Geographical barriers can separate populations of one species to prevent interbreeding. As explained in the next answer, this is important in speciation. Some features that act as geographical barriers are oceans, rivers, mountain ranges, deserts and deep canyons.

4-37 A single population of one species contains **genetically determined variation**. A **geographical barrier** splits the original population into two. The isolated populations are subjected to **different environmental conditions**. **Different phenotypes** are selected in the two isolated populations. Over a number of **generations,** the populations become different. **Mutations** occur that affect reproductive structures and behaviours. If after removal of the geographical barrier **interbreeding is no longer possible,** then **two species** have formed.

4-38 To determine whether one or two species are present, you need to look for interbreeding. If individuals from the two groups can mate to produce vigorous fertile offspring under natural conditions, they are members of one species. If interbreeding or gene flow is not possible, two species are present. Remember that domestic dogs belong to the one species even though the largest and the smallest do not mate. They do mate with dogs closer to their size and so gene flow is possible.

4-39 Two types of isolating mechanisms exist. Pre-mating isolating mechanisms prevent successful mating. Post-mating mechanisms prevent successful production of vigorous and fertile offspring.

Pre-mating isolating mechanisms include habitat isolation, seasonal isolation, differences in courtship, and mechanical isolation (reproductive organs are not compatible).

Post-mating isolating mechanisms include gamete mortality (gametes do not survive until fertilisation), zygote mortality (formed but not viable), and hybrid sterility (for example, mules).

4-40 There is no major difference between 'Natural' and 'Selective Breeding'. They both produce evolutionary change. The difference is the selecting agent. In 'Natural Selection', it is the environment at large. In the case of 'Artificial Selection', it could be a dairy farmer keeping certain cows and removing others from the herd.

4-41 Selective breeding reduces genetic diversity. The advantage of this is that the farmer produces more individuals with the desired trait. It is more efficient. The disadvantage is that if the environment changes to the detriment of one individual then as all individuals are genetically similar, it will be to the detriment of all and the crop will fail.

4-42 Biodiversity is the variation shown by all living things on Earth.

4-43 4500 million years old

4-44 Early Earth consisted of:
- massive oceans with few landmasses
- no ozone layer therefore very high level of ultraviolet radiation
- atmosphere contained high levels of water vapour, carbon dioxide, and nitrogen with lower levels of methane, ammonia, hydrogen sulfide and sulfur dioxide. No oxygen.
- volcanic activity added heat, gases, ash and dust to the atmosphere
- many electrical storms
- no life.

4-45 Around 3800 million years ago the atmosphere became thicker due to the addition of gases from volcanic activity. This increased atmospheric pressure and resulted in condensation of water vapour into liquid water. Evidence for this comes from dating sedimentary rocks. Sedimentary rocks are formed under water. The oldest sedimentary rocks have been dated at around 3800 million years old. Ripple marks on ancient rocks add further support.

4-46 Organic molecules are molecules produced by living organisms. They contain carbon, hydrogen and oxygen. They include carbohydrates, lipids, proteins and nucleic acids. (By definition not included are; carbon dioxide, oxygen or water.)

4-47 The presence of carbon, oxygen, hydrogen, nitrogen, phosphorous and sulphur were necessary before life could form.

4-48 Two other factors essential for life on Earth were
- liquid water. A solvent in which the chemical reactions essential to life could occur
- energy source. Living organisms require energy to maintain structure, grow and function.

4-49 Urey and Miller recreated in the laboratory the conditions (reducing –methane, ammonia, hydrogen and water and energy-rich –electric discharge) thought to be present on early Earth. They found that organic molecules formed. This was significant as it showed that organic molecules and therefore life could have originated on Earth.

4-50 Oxygen readily reacts with organic molecules. If oxygen had been present in the atmosphere of early Earth the newly formed small organic molecules would have been destroyed before they could form larger molecules.

4-51 Steps required:
- the formation of small organic molecules
- the formation of organic polymers from the small organic molecules
- the formation of self-replicating molecules, e.g. DNA
- the packaging of organic molecules in a membrane.

4-52 Approximately 3850 MYR ago

4-53

Major stage in the evolution of life	Description
1 Formation of organic molecules	Complex organic molecules, e.g. amino acids formed in water
2 Formation of membranes	A membrane formed around the organic molecules isolating them from the external environment
3 Formation of prokaryotic, anaerobic, heterotrophic cells (about 3850 MYR)	The membrane bound cells lacked membrane bound organelles, were heterotrophic, i.e. could absorb organic molecules across the membrane and respired anaerobically as there was little oxygen in the atmosphere
4 Formation of prokaryotic autotrophic cells (about 3400 MYR)	Some prokaryotes were able to produce their own organic material from inorganic molecules using either the energy from sunlight (photosynthesis) or the energy from simple chemical reactions (chemosynthesis). Oxygen was produced as product in photosynthesis. This increased the amount of oxygen in the atmosphere
5 Formation of aerobic prokaryotic cells (about 2000 MYR)	Photosynthetic prokaryotic cells added oxygen to the atmosphere. Oxygen is required for aerobic respiration, which is more efficient than aerobic respiration.
6 Formation of eukaryotic cells (about 1800 MYR)	Cells with membrane-bound organelles possibly formed by infolding of the outer membrane or endosymbiosis between prokaryotes
7 Formation of colonial organisms	May have been formed from incomplete cell division/separation where a number of individuals were held together in a jelly
8 Formation of multicellular organism (about 1500 MYR)	The organism consists of a number of different types of cells each performing a specialised function

4-54 Palaeontological evidence: 3.5 billion-year-old fossil stromatolites (formed by prokaryotes) have been found in Western Australia and Southern Africa.

4-55 Geological evidence includes: chemical analysis of 3.850 billion-year-old rocks indicate a higher ratio of ^{12}C to ^{13}C. Living organisms use ^{12}C in preference to ^{13}C in chemical reactions. A higher ratio suggests that ^{12}C has been added by living organisms.

4-56 **When**: oxygen was added to the atmosphere about 2000 MYR ago.
How: oxygen is produced in photosynthesis and was added to the atmosphere by photosynthetic microbes. Most scientist think that at first the oxygen produced (3400 MYR ago) reacted with ferrous ions present in sea water to form ferric oxide which settled in layers on the sea floor. After all the dissolve ferrous ions in the sea had reacted (2000 MYR ago), the oxygen produced in photosynthesis was added to the atmosphere.
Evidence: The presence and age of
- rocks containing banded iron formation (layers of ferric oxide)
- red beds formed by oxygen reacting with mineral deposits on land
- aerobic microbe fossils

4-57 The endosymbiotic theory of eukaryotic evolution: Chloroplasts and mitochondria where once free-living prokaryotes. They were engulfed by a larger cell forming a mutually beneficial partnership.
Evidence:
- mitochondria and chloroplasts contain their own DNA and this DNA similar to that found in prokaryotes
- mitochondria and chloroplasts contain their own ribosomes and these are similar to ribosomes found in prokaryotes
- mitochondria and chloroplasts are self-replicating.

4-58 Aerobic respiration is more efficient as more energy per glucose molecule is released to the cell than anaerobic respiration. This means that there is more energy available to the cell for active processes.

4-59 In the upper atmosphere, sunlight split oxygen molecules into oxygen atoms. These reacted with other oxygen molecules to form ozone (O_3). The ozone layer absorbs UV radiation and made it possible for organisms to live on land.

4-60

First flowering plants
First mammals
First conifers
Earliest reptiles
First ferns
First amphibians
First land plants
First insects
First vertebrates
First multicellular eukaryotes
First unicellular eukaryotes
First aerobic microbes
First photosynthetic bacteria
First prokaryotic cells
Origin of the Earth
0
500
1000
1500
2000
2500
3000
3500
4000
4500

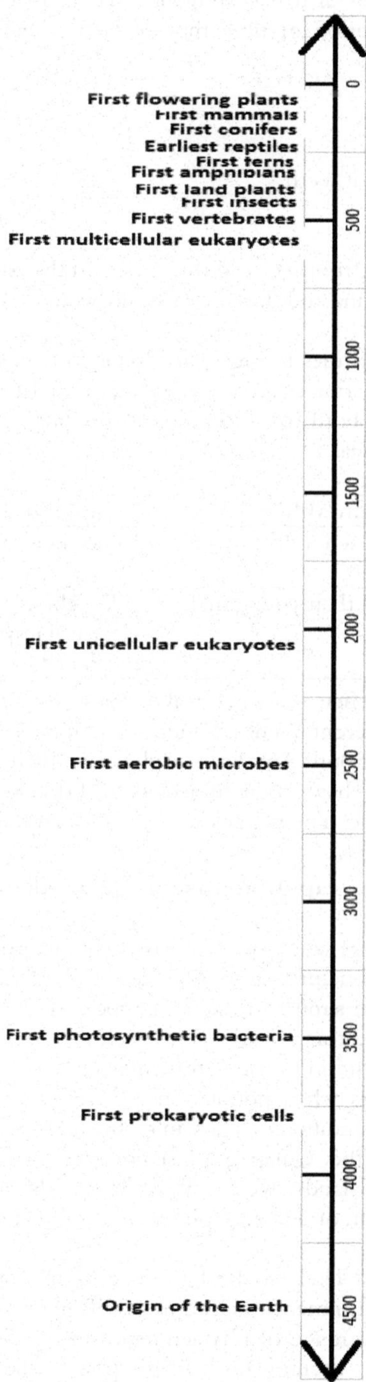

4-61 A transitional series is a series of fossils showing small changes from one form to another and after a long period of time may result is organisms that are very different from their ancestors.

4-62 The ancestor of horses was
- wombat sized
- ate soft leaves and fruit
- had generalised and fairly flat teeth
- had shorter legs
- had five toes on each foot
- walked on four toes on the front feet and three toes on the back feet.

4-63 Darwin observed that the fauna and flora in Australia were:
- unique, e.g. monotremes
- Whilst different show similarities to organisms living in the same environment in different parts of the world. For example, flying squirrels of North America and glider possums of Australia both use skin flaps for gliding yet one is a placental mammal and one is a marsupial.

4-64 Platypus features

Reptile-like	Mammal-like	Unique to the platypus
Lay soft, leathery eggs	Covered in hair	Venomous spur
Presence of a cloaca	Feed their young milk	Bill – ability to detect electrical impulses produced by prey

4-65 Adaptive radiation occurs when one ancestral species gives rise to a number of new species occupying different niches. Finches on the Galapagos Islands are descended from a mainland South American seed-eating finch species. The finches live on different islands and have been subjected to different selection pressures such as type of food available. The finches now vary in what they eat and the size and shape of their beaks.

4-66 Adaptive radiation involves a **rapid increase** in species diversity within a **short time**.

4-67 **Convergent evolution** occurs when two unrelated species under similar selective pressures develop similar structures and appearance. For example, the placental mole and marsupial mole are similar in size and appearance. They have, however, evolved from different ancestors. Convergent evolution occurs when unrelated species develop the same solution to **the same** environmental problem.
Divergent evolution occurs when populations of one species are subjected to different environmental conditions. Darwin's finches provide an excellent example. Colonisation of the Galapagos Islands was by one finch species. Populations on different islands were subjected to different environmental pressures. There was selection of different phenotypes on different islands.

4-68 Adaptive radiation.

4-69 Divergent evolution is most likely to lead to speciation. Populations from one species live in different environments. Selection of different phenotypes occurs in different environments. The range of characteristics in the different populations increases. Differences may accumulate to the point where interbreeding is impossible even if members of different populations are brought together.

4-70 Individuals cannot become extinct. Individuals **die**. If all individuals of a species are dead, the species is **extinct**. A species that has living representatives is **extant**.

4-71 The punctuated equilibrium model of evolution states that evolution occurs in fits and starts. Periods of rapid evolutionary change are followed by long periods of little evolutionary change. When rapid change occurs, organisms either die out or move to other areas. This leads to the development of small isolated pockets of organisms. Evolutionary change is faster in small isolated populations than large populations.

4-72 Punctuated equilibrium involves rapid evolution whereas natural selection involves slow change. Long periods of very little change are a characteristic of punctuated equilibrium whereas change is gradual in natural select ion.

4-73

	Evidence for evolution	**Example**
Biochemical (molecular homology)	The biochemistry of related organisms is similar. If new species have evolved from a common ancestor, it would be expected that they would have similar molecules and biochemical pathways. The biochemistry of an individual is determined by the proteins/enzymes it makes and these are determined by the organism's DNA. Closely related organisms have had less time for mutations to accumulate and are therefore more similar.	**-similar DNA/proteins in closely related individuals**
Comparative anatomy -homologous structures	If new species have evolved from a common ancestor, it would be expected that the new and ancestral species would have the same basic structure. The basic structure may be modified according to the organism's way of life	The pentadactyl limb: All mammals have similar structures. The skeleton is similar. The same bone pattern is present although the size and shape of bones may vary.
-vestigial organs	If new species have evolved from a common ancestor, structures may be present that no longer have a function. These structures would have had a function in ancestral organisms.	- presence of an appendix in humans. - vestigial hind limbs in whales.
Comparative embryology	The developmental stages of related organisms are similar. If new species have evolved from a common ancestor, it would be expected that they would have similar stages in their development.	All chordates (including fish, amphibians, birds and mammals) have gills slits at some stage in their life cycle.

Biogeography	By studying the distribution of organisms today and comparing them to fossil distribution and earth movements. If new species have evolved from a common ancestor, then species found in isolated areas will be more similar to species that lived in the same area in the past than to species that live in distant areas with the same environment today.	-The finches on the Galapagos Islands are more similar to each other than finches found in other places. This supports the idea that the Galapagos finches evolved from a small group of ancestral finches. -The Australian continent separated from the continents before the rise of placental mammals. For this reason, Australia was populated by a wide range of marsupials that filled all ecological niches.
Palaeontology **- transitional forms**	Fossil record shows that living organisms have changed over time. If new species have evolved from a common ancestor, then the fossil record should include fossils that have characteristics that are intermediate between the ancestral form and the new form.	- ferns appear in the fossil record before gymnosperms and angiosperms appear most recently. - *Archaeopteryx* fossils: intermediate between birds and reptiles - Lung fish: intermediate between fish and amphibians

4-74 Molecular homology involves looking at the similarity between molecules found in different organisms. DNA and amino acid sequences are often used in molecular homology. The genetic code is universal. This suggests that all living things are related. Differences in nucleotide sequence arise by mutation. The greater the number of mutations present the longer the time since divergence from a common ancestor. As DNA codes for the sequence of amino acid in a protein, the order of amino acids is also an indication of the time that has passed since divergence.

4-75 Comparison of mitochondrial DNA and nuclear DNA.

	Mitochondrial DNA	**Nuclear DNA**
Shape	Circular	Linear
Location	Within the mitochondria	Within the nucleus
Inherited from	Mother	Mother and father
Number of alleles per gene per cell	Many	Two alleles per gene. The alleles maybe the same or different.

4-76 When a sperm fertilises an ovum, the sperm's nucleus passes into the ovum. This forms the first cell of the new individual and therefore the mitochondrial DNA present, comes from the mother not the father. This makes mitochondrial DNA very useful in the study of evolution as changes in mitochondrial DNA are due to mutation and not recombination during meiosis. Evolutionary change can be measured by comparing the mitochondrial DNA from two different species. The longer the time passed since their divergence, the longer the time for the accumulation of different mutations.

4-77 In DNA hybridisation, the DNA of two different organisms are mixed, denatured and allowed to join together. The DNA formed consists of one chain of nucleotides from each organism. The newly formed DNA is heated again. The more similar the DNA of the organisms, the higher the temperature required to separate the nucleotide chains. Organisms that have a recent common ancestor will have more similar DNA than organisms who have an older evolutionary relationship.

4-78

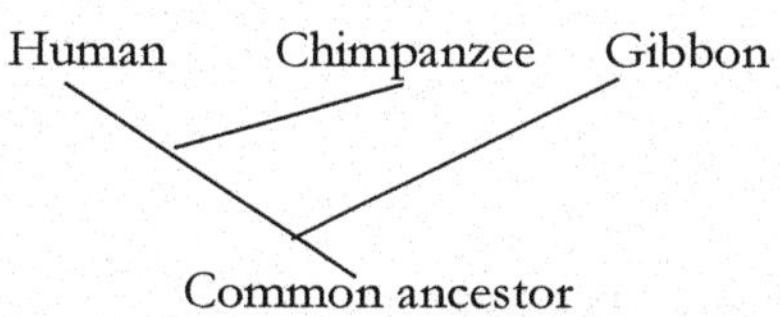

4-79 Analogous structures are not inherited from a common ancestor whereas homologous structures are. As a result, homologous structures have the same basic structure whereas the basic structure of analogous structures is not always similar. Analogous structures have similar functions, but this is not always true of homologous structures. Analogous structures have been subject to the same selection pressure.

4-80 The Theory of Continental Drift: The present-day continents were once one large land mass that split into pieces that have and continue to move apart.

4-81 The Theory of Continental drift states that the continents are not fixed, rather they once formed one large land mass that split into pieces that have and continue to move apart. There was no explanation of how the continents move. Hesse suggested the continents move because they are attached to large plates of crust that float on a hot underlying mantle. Plates may move apart, converge or slide pass each other. Where they move apart hot material from the mantle rises, cools and adds to the plate (spreading sea floor). Where plates converge earthquakes and volcanoes occur and mountain ranges and trenches form.

4-82 Asia, Africa, North America, South America, Europe and Australia

4-83 Pangaea

4-84 About 180 million years ag.

4-85 North America, Europe and Asia (except for India)

4-86

Millions of years ago	
	Gondwana – South America/Africa/Madagascar/India/
150	Antarctica/Australia/New Zealand
140	Africa and South America separated
	India moved north
120	
100	
80	Madagascar split from Africa
	New Zealand split from Australia
60	South America and Australia/Antarctic separate
40	Australia/New Guinea and Antarctic separate
20	
Present	

4-87 Australia is moving north at approximately 6 cm per year.

4-88

Evidence	**Description**
Matching continental margins -shape -type of rock	The shape of the southern continents fit together like a jigsaw. Where the pieces would have met there is **similar geology** (matching tillite-coal layers, dolerite formations in Tasmania and Antarctica, basalt in South America and Africa) suggesting that these areas were formed in the **same place** and at the **same time**.
Position mid-ocean ridges	Mid ocean ridges have been found as predicted where plates are moving apart.
Spreading zones between continental plates	The ocean floor rock increases in age the further it is from the mid-ocean ridges. This indicates that the plates are moving apart, and new rock is forming between them.
Fossils in common on Gondwanan continents	Presence of the **same type** and **age** fossils indicate the same organisms lived at the same time on Gondwanan continents. As many of these could not disperse across oceans, the continents must have been joined at one time. For example, plant fossils *Glossopteris* and *Gangamopteris* have been found in rocks of all Gondwanan continents and nowhere else.

	Fossils of the reptile *Lystrosaurus* has been found in Africa, India and Antarctica. Platypus fossils have been found in Australia and South America. The earliest fossils of marsupials have been found in North America. Some scientists suggest that marsupials migrated to Europe and South America when they were part of Pangaea. When Pangaea broke up those marsupials in Laurasia died out, those in South America migrated to Antarctica and then Australia. The lack of marsupials in New Zealand indicates that New Zealand broke off from Gondwana before the marsupials had reached New Zealand.
Similarities between present day organisms on Gondwanan continents	Many present-day animals and plants occurring on the Gondwanan continents share a common ancestor that existed before Gondwana broke apart. Their present-day distribution is evidence. For example, *Nothofagus*) is found today in New Zealand, Australia, New Guinea and South America. (Fossil pollen has been found in Antarctica.) This indicates that *Nothofagus* evolved after South Africa broke off from Gondwana. The ratites (flightless birds) occur only in the southern hemisphere.

4-89 Megafauna: giant extinct vertebrates that were much larger than their present-day relatives.

4-90 Australia's megafauna became extinct about 50 000–20 000 years ago. Two theories for the extinction include:
- overhunting or habitat destruction by early indigenous Australians
- climate change.

4-91 An extant species is a present-day species.

4-92 A living fossil is a present-day species that has changed little across its fossil record. Examples include: stromatolites, *Nothofagus*, the Wollemi pine, cycads, velvet-worms, and crocodiles.

4-93 Fossils provide evidence of once-living organisms. Fossils are not always the dead organism's remains. Fossil footprints or body impressions can also provide evidence that an organism existed. Sometimes the actual remains may no longer exist. A cast showing size, shape and texture of the outer surface may remain. Fossils are not always the dead remains but can be evidence that an organism was in the area.

4-94 Any condition that reduces scavenging and decay by other organisms will assist fossilisation.

– **Quick burial**: Fossils are most likely to occur in an environment where there is rapid burial of the dead organism. This reduces the exposure to scavengers and decomposers. Burial of marine organisms in sediments after death is more likely than burial in a terrestrial environment. Quick burial of an organism by volcanic ash provides excellent conditions for fossilisation. A visit to Pompeii in Italy is well worthwhile. The entire town was buried in about 79 A.D. when Vesuvius erupted. Casts of the residents have been obtained by pouring plaster of Paris into spaces in the compressed ash.

– **Cold temperatures**: Frozen organisms are fossils. Low temperatures protect the remains from decomposers. Consider woolly mammoth remains!
– **Sap**: Sap also protects organisms from decay. Insects in amber were a source of DNA in the movie *Jurassic Park* and its sequels.
– **Anoxic environments**: The remains are protected, as aerobic decomposers cannot survive in anoxic environments.
Disturbance of organisms after death reduces chances of fossilisation. Exposure to wind, rain, scavengers and bacteria also impede fossilisation. Earth movements and erosion after burial tend to damage fossils. Erosion of sedimentary rocks, however, is the major way fossils are revealed allowing people to find them.

4-95 Evidence of the once-living organism's environment can be deduced by looking at the sediments in which it is buried and at the other fossils present. An organism whose fossil is found in sandstone probably lived in a marine environment, whereas the fossil of an organism living by a lake would probably be found in mudstone. Fossilised pollen and seeds can indicate the types of plants growing in an area in the past. The condition of a fossilised skeleton can tell a story. Carnivores are likely to have torn apart the skeleton of smaller animals. Fossilised teeth can indicate the diet and age of the organism before it died. The teeth of carnivores and herbivores are very different. Tooth wear indicates the age of an organism at death. For a toothed animal, the greater the teeth wear, the older the animal was at death.

4-96 Fossils are difficult to identify to the species level for two major reasons:
– The remains are incomplete, and so it is difficult to compare relationships with living organisms.
– A major aspect of species identification relies on determining whether two individuals can interbreed to produce fertile vigorous offspring. So far, there have been no reports of fossils mating to produce offspring! That fossils are dead may present some insurmountable problems to conducting an active sex life!

4-97 Absolute dating gives an estimate of age in years. For example, one million + or – ten thousand years. Absolute dates are calculated using the known rates of decay of radioactive isotopes.
Relative dating relies on sedimentary layers (strata) being laid down in order. In an environment undisturbed by earth movement, the oldest fossils will be buried more deeply than younger fossils. An actual date cannot be determined with relative dating.

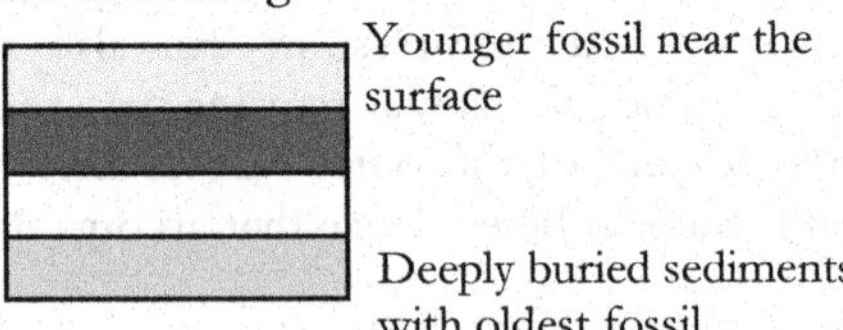

4-98 Radio-isotopic dating relies on palaeontologists knowing the half-life of the isotope with which they are dealing. Radioactive isotopes break down at known rates. Carbon-14's half-life is 5568 years. If you had 1 gram of carbon-14 today, in 5568 years you would have 0.5 grams. After another 5560 years (a total of 11 136 years) then you would have 0.25 grams of carbon-14. Radioactive rubidium isotopes have a half-life of 47 000 million years.

4-99 **Carbon-14**: This method is used for dating organic matter up to 50 000 years in age. Note: Carbon occurs in all organic molecules, therefore the ^{14}C isotope is useful for dating organic matter.

Potassium/Argon: Used to date volcanic rock from 10 000 to 100 million years old.

Fission Track: This method can be used to date volcanic rock over 10 000 years old.

Uranium/Thorium: A method used to date bone and teeth of less than 350 000 years old.

Direct dating of fossils is not always possible. If sediment formation was at the same time as the fossilisation, dating the sediment dates the fossil.

4-100 Palaeontologists believe *Archaeopteryx* represents an intermediate stage in the evolution of birds from reptile-like ancestors. A problem for evolutionists is the lack of a complete fossil record. The fossil record has many gaps. Excitement is generated by fossils, like *Archaeopteryx*, because they fill gaps in the fossil record.

4-101 Cane toads at the expansive front are:
- larger
- legs are longer and stronger
- move further and in a straight line
- breed faster.

4-102 Different selection pressures act on the cane toads depending on their position. It is an advantage for those on the expansive front to move quickly to reduce completion with other cane toads. Speed is not an advantage for those cane toads back from the expansive front as there is no way they can move away from other cane toads.

4-103 There was original variation in DDT resistance in mosquitoes. Some were resistant, and others were not. In areas where DDT was sprayed, those mosquitoes that were not resistant died. They produced less offspring. Those mosquitoes that were resistant to DDT survived, reproduced and their offspring inherited DDT resistance. Over time, more individuals in the population have become resistant to DDT.

Section I Questions

4-104 All of the following are abiotic factors except

(C) **humus content of soil.** (Humus consists of organic remains of organisms.)

4-105 All of the following are biotic factors except

(B) **the concentration phosphates in the soil.** (Phosphate is an inorganic nutrient).

4-106 Abiotic factors that affect living organisms in the marine environment are

(B) **salinity of the water.** (Salinity is a measure of the salt concentration, an inorganic substance found in seawater. Phytoplankton are photosynthetic organisms, ammonia is a by-product of fish activity so will be related to numbers of fish and broadcast spawning refers to the release of eggs and sperm at one time – all of these are biotic factors.)

4-107 In comparing problems of living on land to living in water, which of the following is correct?

(B) **water offers more support than air to aquatic animals.** (This is a fact. Other alternatives are incorrect – there is less oxygen available in water, light does penetrate the ocean deeply and desiccation (drying out) is not a problem.

4-108 The cane toad was introduced into Australia as a biological control for cane beetles. Control of cane beetles was unsuccessful because

(C) cane toads cannot jump very high, so the cane beetles were able to retreat to the top of the sugar cane. (Cane toads rapidly reproduced and could digest cane beetles but the cane beetles were unaffected because the cane toads could not catch them!)

4-109 Cane toads are not a problem in Hawaii and South America because

(D) predators are present (Food and water are available and climatic conditions are similar to parts of Australia. The presence of predators limits the population.)

4-110 The selection pressure in this example is:

(A) the presence of *Cactoblastis cactorum*. (The selection pressure is the factor in the environment that determines which organisms survive.)

4-111 Structural adaptations include the following

(C) tough waterproof eggshell to protect bird embryo. (The egg shell is a structure. The other alternatives describe physiology or behaviour.)

4-112 Behavioural adaptations include the following:

(D) the desert hopping mouse's nocturnal habits. (The other alternatives describe physiology or structure.)

4-113 Shelled eggs are an adaptation to living on land because they

(D) they prevent dehydration of the embryo. (The other alternatives are true but are not directly related to living on land.)

4-114 Plant leaves of different species have features that allow them to survive in their particular environment. It is reasonable to suggest that

(B) the hanging leaves of gum trees reduces water loss. (Less light and heat fall on the hanging leaf – **this** reduces water loss by transpiration.)

4-115 A plant adaptation commonly found in dry climates is

(C) plants with an inverted stomatal rhythm. (Stomata open at night or cooler times to reduce water loss.)

4-116 Desert animals, such as the desert hopping mouse, have a variety of adaptations to overcome a lack of regular water supply such as

(D) greater reliance on the use of metabolic water. (Many desert animals are able to use efficiently water produced during cellular respiration.)

4-117 An animal's body can gain or lose heat depending on the temperature difference between the animal's body temperature and the temperature of the environment. One of the following is correct.

(C) animals can gain or lose heat by convection. (Heat is only lost by evaporation and moist efficiently in dry conditions. Heat will be lost if the surrounds are colder. Heat can be gained, or lost, by convection.)

4-118 In cold environments, animals can increase heat production by

(D) increasing their rate of shivering. (Fact, none of the others are increasing heat production.)

4-119 In hot environments animals can increase heat loss by

(A) increasing blood flow to the body's extremities (e.g. tips of ears). (Heat is lost by radiation from the extremities.)

4-120 Which of the following populations would be less likely to survive a change in environment?

(C) a population of individuals showing many phenotypic variations. (Populations with the greatest variation are more likely to survive environmental change. Asexually reproducing bacteria are genetically and therefore phenotypically similar. Populations descended from a few individuals will be genetically and phenotypically similar.)

4-121 Biodiversity refers to

(D) the variety of all life on Earth. (Definition.)

4-122 Biodiversity has been decreased by

(D) all of the above. (Clearing of forests, global warming and agriculture have all resulted in a decrease in biodiversity.)

4-123 The age of the Earth is has been estimated by scientists as approximately

(C) 4500 million years old. (Fact)

4-124 The early atmosphere of the Earth contained

(B) high levels of carbon dioxide. (There was no free oxygen, no ozone layer and low levels of ammonia in the primeval atmosphere of the Earth.)

4-125 The scientists Alexander Oparin and John Haldane were the first to

(B) suggest the chemicals for life came from Earth (Svante Arrhenius suggested that life came from outer space and Harold Urey worked with Stanley Miller to demonstrate that inorganic molecules could be converted into organic molecules)

4-126 Organic molecules contain

(B) carbon, hydrogen and oxygen. (Ozone is not an element, iron is not found in all organic molecules, but carbon is a component of all organic compounds.)

4-127 The experiments of Harold Urey and Stanley Miller

(D) showed that life on Earth could have originated on Earth. (Urey and Miller performed experiments where an electric current was passed through conditions thought to be like that of early Earth. Organic molecules formed. This supports but does not prove that life originated on Earth. The experiments show that is it possible that life may have originated on Earth.)

4-128 Harold Urey and Stanley Miller recreated the atmosphere of early Earth in the laboratory and produced

(A) amino acids. (Proteins are made of amino acids, but they were not formed in the experiment. Nucleic acids and fatty acids were also not formed.

4-129 Conditions required for the formation of organic molecules in the laboratory include

(C) electric discharge. (Oxygen reacts with the organic molecules as they form, light is not required but heat is.)

4-130 Which of the following shows the generally accepted order beginning with the oldest, of the major stages in the evolution of living things?

(B) anaerobic prokaryotes, photosynthetic prokaryotes, aerobic prokaryotes, eukaryotes. (Prokaryotes occurred before eukaryotes. The atmosphere of early earth had little oxygen therefore anaerobic prokaryote were first. Photosynthetic prokaryotes added oxygen to the atmosphere which could then be used by aerobic prokaryotes.)

4-131 What was the energy source for the first forms of life?

(B) ingestion of simple organic molecules from their surroundings (It is thought that the first forms of life were anaerobic heterotrophs.)

4-132 The endosymbiotic theory suggests

(A) mitochondria were once free-living prokaryotes. (The endosymbiotic theory of eukaryotic evolution: Chloroplasts and mitochondria where once free-living prokaryotes. They were engulfed by a larger cell forming a mutually beneficial partnership.)

4-133 The correct order from first to last is

(D) ii, iv, i, iii (Simple organic molecules would need to be formed before they could be assembled into organic polymers. Self-replicating molecules, e.g. DNA would lead to the production of enzymes and the possibility of chemical pathways to sustain life and finally the packaging of organic molecules in a membrane creating the first cell.)

4-134 The atmosphere of early Earth contained very little free oxygen. Gradually oxygen was added to the atmosphere by

(C) photosynthetic autotrophs. (Oxygen is produced in photosynthesis.)

4-135 The following is an example of divergent evolution:

(C) Australian marsupials present in rainforests, grasslands and open woodlands. (Divergent evolution results in organisms that appear different because of different selection pressures but have a recent common ancestor.)

4-136 The different size and shape of the beaks of different species of finches on the Galapagos Islands is an example of adaptive radiation. The selection pressure involved is

(A) the type of food source. (The selection pressure is the environmental factor that determines which individuals survive. The beaks are different because different food sources are found on each island. The birds with the beak that best enables them to access the food are selected for.)

4-137 Punctuated equilibrium differs from natural selection in that

(D) punctuated equilibrium involves rapid change whereas natural selection involves slow change. (The punctuated equilibrium model of evolution states that evolution occurs in fits and starts. Periods of rapid evolutionary change are followed by long periods of little evolutionary change. Change is gradual when brought about by natural selection.)

4-138 Using this data which animal is most distantly related to humans?

(D) Frog (The more distant the relationship, the longer the time for mutations to accumulate since divergence.)

4-139 Based on this information, it may be concluded that

(A) humans are more closely related to rhesus monkeys than horses. (Humans share a more recent common ancestor/point of divergence with rhesus monkeys than they do with horses.)

4-140 Amphibians, reptiles, birds and mammals have pentadactyl limbs this suggests

(D) they share a common ancestor. (Pentadactyl limbs have five digits and many do not live on land.)

4-141 The name of the supercontinent that formed about 260 million years ago is

(A) Pangaea. (Pangaea split to form Laurasia and Gondwana. Antarctica was part of Gondwana.)

4-142 The following were part of Laurasia

(D) North America, Europe and Asia. (South America, Africa, Australia, Antarctica and India were part of Gondwana.)

4-143 The following were part of Gondwana

(B) South America, Africa and Australia. (North America, Europe and Asia were part of Laurasia.)

4-144 The following will most likely occur where continental plates move apart

(B) formation of new sea floor (Uplifting of mountain ranges occurs where plates collide. Where the plates move apart molten material from the mantle moves to the surface and forms new sea floor.)

4-145 Which of the following provide support for the existence of Gondwana?

(D) the distribution of ratite birds across the southern hemisphere. (Flightless birds occur on the southern landmass. These birds share a recent common ancestor. They cannot fly therefore they must have evolved together and then have been dispersed with Gondwana broke up.)

4-146 Gondwana began to break up with a split between

(A) South America and Africa. (The first split was between South America and Africa, then New Zealand, then South America and finally Australia from Antarctica.)

4-147 The Australian megafauna

(C) may have become extinct as a result of climate change. (The Australian megafauna did not include any placental mammals, e.g. cows, coexisted with indigenous Australians and were well adapted to cooler climates.)

4-148 The flightless ratite birds are only found in the southern hemisphere. It is thought that they shared a common ancestor. Their distribution

(A) is evidence that Gondwana once existed. (They have a common ancestor therefore they have not arisen independently or that flightlessness has been selected for in different places. Laurasia did not include the southern landmasses.)

4-149 Which of the following best describe the vegetation of Gondwana about 150 million years ago?

(A) conifers, cycads, ferns (Flowering plants appeared about 110 million years ago.)

4-150 Sclerophyll plants

(B) have small leaves resulting in reduced water loss. (Sclerophyll plants are adapted to dry, hot environment. They have small leaves to reduce water loss and heat gain.)

4-151 Changes in Australian fauna and flora is thought to be due to

(D) all of the above. (Continental drift has brought about climate change and along with human intervention is thought to have resulted in changing Australian fauna and flora.)

4-152 Supporting palaeontological evidence includes

(C) fossils dating back 3500 million years. (All other alternatives refer to geological evidence. Palaeontological evidence refers to fossil evidence.)

4-153 These diagrams suggest that fossils found

(D) in layers C and E are the same age

4-154 Insects in amber are examples of

(B) direct fossils. (The amber has protected the insects from decomposers therefore some of the original organic matter remains.)

4-155 ^{14}C has a half-life of approximately 6000 years. What percentage of the original ^{14}C would remain in a 12 000-year-old fossil?

(C) 25%. (After 6000 years or 1 half-life, 50% would remain. After another 6000 years, half of 50% or 25% would remain.)

4-156 The oldest dingo fossil has been dated at 3450 years old using

(A) ^{14}C dating of the fossil bones. (Potassium-argon dating is used for very old fossils. ^{14}C is used to date fossils up to about 50 000 years old. The fossils itself is dated not the surrounding rock.)

4-157 Very few fossils of jellyfish have been found because

(D) they are not made of hard structures. (Jellyfish are thought to have evolved a long time ago, living in water would not help fossilisation and many jellyfish are not small. Jellyfish do not contain hard parts. These more readily fossilise.)

Section II Questions

4-158

(a) Possible abiotic factors include: air temperature (thermometer), water temperature (thermometer), salinity (conductivity meter), wind (anemometer), light intensity (light meter) and pH (pH meter).

(b) Predation is the major factor that affects distribution. Predators that live in the water can move in over the platform at high tide. Organisms that can survive periods out of the water can avoid predation by living near the high tide mark.

(c) A steep slope results in narrow zones. A relatively flat platform may have bands or zones that are nearly as wide as the platform. On a steep shore, environmental conditions change quickly so the adaptations needed differ, which results in zones of different organisms.

(d) Permanent rock pools
Cracks in the rock
Shelters under or behind rocks
Shelters under seaweed

(e) Structural adaptation:
- tough exterior (exoskeleton) of crabs and after crustaceans – protect against wave damage.
- strong shell of predatory molluscs – protect against wave damage and drying out.

Behavioural adaptation:
- sheltering in rock pools or cracks at low tide – to reduce water loss.
- movement up the rock platform to feed at high tide – potential food availability.

4-159

(a) The cane toad thrived because
- it was well adapted to the Australian environment
- it eats a wide range of insects and other small animals
- it produces many offspring quickly
- it readily disperses
- lack of predators.

(b) There was original genetic variation that resulted in phenotypic variation in adult head size of red-bellied black snakes. Those red-bellied black snakes that had large heads were able to swallow large toads containing high concentrations of toxins. These snakes died and therefore produced fewer offspring. Those snakes with smaller heads were able to only swallow small toads. They ingested less poison, survived and produced more offspring who inherited a small head. Over many generations the proportion of snakes with small heads increased in the population.

(c) The selection pressure is competition with other cane toads. Large toads with long legs can move quickly and therefore disperse into new areas where there are few competitors.

4-160

(a) The atmosphere of early Earth did not contain free oxygen therefore the early forms of life could not be aerobic and must have been anaerobic.

(b) The obtained energy by ingesting free organic molecules from their surroundings.

(c) Photosynthetic organisms produce oxygen as a product of photosynthesis. This oxygen was released by the plants into the atmosphere resulting in an increase in free atmospheric oxygen.
This leads to
- the development of the ozone layer and this in turn lead to a decrease in UV radiation allowing organisms to live on land.
- organism to obtain energy from aerobic respiration. Oxygen is required for aerobic respiration. Aerobic respiration is more efficient as more energy per glucose molecule is released to the cell. This means that there is more energy available to the cell for active processes.

4-161

(a) Adaptive radiation occurs when an ancestral species is subjected to different selection pressures. Colonisation of the Galapagos Islands was by one finch species. Populations on different island were subjected to different environmental pressures – different food sources. Different beak sizes and shapes were selected for resulting in the range of finches now found on the Galapagos Islands.

(b) Divergent evolution

4-162

(a) Organisms that have a close evolutionary relationship have a recent common ancestor therefore their DNA and the resulting proteins produced will be similar. The more similar the DNA/proteins, the closer the relationship.

(b) Most fossils contain very little organic material. Molecular comparison can only occur if organic matter is present. PCR (Polymerase Chain Reaction) enables small amount of DNA to be amplified and studied.

(c)

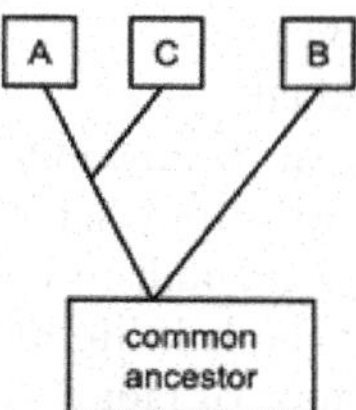

(d) It is assumed that organisms that have similar structures have a close evolutionary relationship however this is not always true. Similar characteristics in very different organisms that are living in similar environments may be selected for. For example, dolphins and sharks have a similar shape. One is a fish and the other a mammal. They do not have a close evolutionary relationship but look similar.

4-163

(a) Gondwana was a large landmass made up of present day South America, Madagascar, India, Antarctica, Australia and New Zealand. When these landmasses were joined their fauna and flora would be similar. Fossil evidence is evidence of past life. The platypus fossils show that the platypus has been present in Australia for 110 million years. During this time the platypus was present in South America therefore South America and Australia must have been joined via the Antarctica. The fossil evidence does support the existence of Gondwana.

(b) The platypus evolved in Australia as this is where the oldest fossils have been found. The platypus moved to South America via the Antarctica when Australia, South America and Antarctica were part of Gondwana.

(c) Any two of the following:
- Africa broke away from Gondwana before the platypus evolved.
- The formation of fossils is a rare event and therefore they have not been found yet.
- Conditions in Africa did not favour the formation of fossils.

4-164

(a) The term 'megafauna' refers to the giant extinct vertebrates that were much larger than their present-day relatives. Examples: *Diprotodon* – a giant wombat, *Procoptodon* – a giant kangaroo, *Thylacoleo carnifex* – a marsupial lion

(b) Human action that could have led to the extinction of the Australian megafauna includes: direct hunting and habitat destruction by fire farming used by indigenous Australians.

(c) Evidence for: extinction occurred about the time indigenous Australians arrived in Australia.
Evidence against: Indigenous Australians and megafauna existed together for many years – kangaroos survived hunting by indigenous Australians and European settlers or no evidence of tools that enabled indigenous Australians to easily kill megafauna.

4-165

(a) Any two of the following reasons:
- the fossils have been undisturbed
- the cave has protected the fossils from wind erosion
- the cave has protected the fossils from water erosion
- the dead organism's remains were undisturbed by predators or scavengers.

(b) Radioisotopic dating where, if the half-life of a radioactive isotope is known, the age can be calculated by the ratios of radioisotope to stable product. For example, any of the following methods:
- potassium/argon
- argon/argon
- lead/lead
- uranium/lead
- rubidium/strontium
- thorium/lead

(No carbon dating as the fossil is more than 50 000 years old.)

(c) Scientists would need to compare the fossil remains with the skeleton of wedge-tailed eagles.

(d) Layer C as it is the deepest.

4-166

(a) Radiometric dating where, if the half-life of a radioactive isotope is known, the age can be calculated by the ratio of radioactive isotope to stable product, e.g. potassium/argon or uranium/lead (not ^{14}C as the fossil is too old).

(b) By comparing the fossil bones with similar bones from living organisms. Same length bones will give an indication of similar size and similar size of muscle attachment sites will give an indication of similar weight.

(c) Any one of the following:
- Only a few Diprotodons lived in Northern Australia as they were not adapted to the northern Australian environment
- Conditions in northern Australia did not favour the fossilisation of Diprotodon bones.

(d) The broken teeth of predator lizards suggest that the lizards killed the diprotodon.

Chapter 5: Ecosystem Dynamics
Content Review Questions – Suggested Answers

5-1 Habitat refers to the living place of an organism. For example, a koala's habitat is a gum tree and a crab might live in a rock pool – its habitat. A community refers to the interacting populations of an area. An ecosystem consists of communities of organisms in their habitats. It refers to the combination of biotic (living) and abiotic (physical or non-living factors) components of a living system.

5-2 The term 'habitat' refers to the place where an organism lives. Microhabitat refers to a more localised or smaller area of the habitat. For example, a crab's habitat might be a rocky shore, but it spends most of its time in cracks on the shore – its microhabitat.

5-3 Some species have very specific requirements or narrow tolerance ranges and so are found within a narrow range. Other species have less specific requirements and may be found over a wider area. For example, a koala that needs specific types of gum leaves compared to rabbits, which have become very common over most of Australia.

5-4 Tolerance range concept.

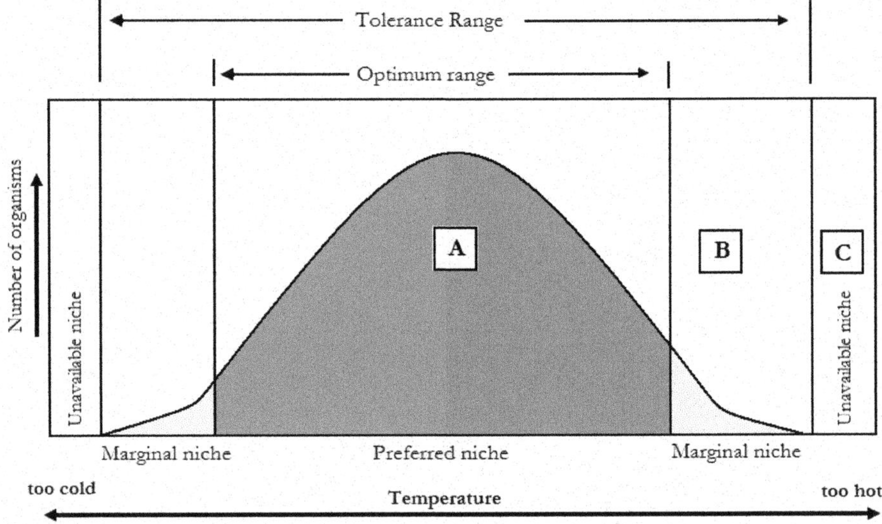

B is the area of physiological stress and **C** is the zone of intolerance.

5-5 A population refers to the number of individuals of one species living in one area, at a particular time. A community refers to the number of interacting populations of different species living in one area. Over time populations change in size and communities may change in terms of species composition.

5-6 Exponential growth as demonstrated in the previous graph which shows at first a slow increase in numbers. As time progresses the slope of the graph increases indicating faster growth.

5-7 Carrying capacity is determined by the resources available to support a population of organisms. This will be influenced by a combination of food resources, suitable habitat and inter-specific competition. Carrying capacity is the population level at which a population can be sustained in a particular area. As a population increases, environmental resistance due to the above density dependent factors will increase, tending to slow further population growth.

5-8 Uncontrolled exponential growth usually leads to population collapse and degradation of the environment that causes a reduction of carrying capacity. The history of the rabbit in Australia is an excellent example. In many areas, rabbits have been in plague proportions. In areas of infestation, they will eat all the available food and then many starve. In the process, their activity reduces food supplies for other herbivores and leads to degradation of the land.

5-9 Intraspecific interactions involve interactions within a species, for example the individuals of the same species of snake competing for mice as a food source. Interspecific interactions involve interactions between different species, for example interactions between the snake and the mouse.

5-10 Rabbits will compete with each other for: food (grass and shrub foliage), mates, and suitable places for burrows. They will compete with other species food (grass and shrub foliage) and other animals that burrow. Rabbits will be prey for carnivores such as foxes, dingoes and wild cats. Rabbits will be susceptible to disease (calicivirus and myxomatosis) and infestation by parasites such as the flea.

5-11 Competitive interactions in the school community include; competing for queue spots at cafeteria, competition for seats at lunch-time, competition for the teacher's attention in class and competition to get the best part in the school play.

5-12 Allelopathy is when an organism (usually plant) produces a chemical that influences (harms or benefits) the growth, survival or reproduction of another organism. This benefits the organism as it reduces competition. For example, pine needles contain allelochemicals that stop the growth of other plants. This reduces competition for water, mineral ions and light between the parent plant and other plants.

5-13 In predator/prey relationships, the predator kills and consumes the prey. In a parasite/host relationship, the host obtains nutrients from its host, but it is in the interest of the parasite not to harm its host too much. Killing its host would mean having to find another host and for internal parasites probable death.

5-14 Factors that affect the numbers in predator and prey populations include:
- the size of the ecosystem
- availability of food/other resources for the prey
- reproductive cycles of both prey and predator
- competition with other predators
- presence of alternative prey.

Note: there are usually more prey than predators.

5-15 Symbiotic interactions refer to interactions between different species where at least one species benefit. Predator/prey relationships are not considered symbiotic.

5-16

Type of symbiosis	**Who benefits/loses**	**Example**
Mutualism	Both	Lichen
Commensalism	One benefits Other not affected	Shark and remora
Parasitism/infectious disease	One benefits Other harmed	Malaria
Amensalism	No benefit to one Other harmed	Bird droppings killing water plants

5-17 The relationship between zooxanthellae and the coral is mutualism. The zooxanthellae produce sugars via photosynthesis. Some sugar is absorbed by the coral polyp – a benefit for the polyp. The zooxanthellae receive shelter from the host coral polyps.

5-18 Most biologists describe the remora/shark relationship as commensalism. One party the remora gets a ride, reducing its need to swim. For the shark if it is not slowed down by the remora there is no downside! One could possibly argue that if the sharks speed is reduced by the presence of remora there is a disadvantage for the shark and the relationship would be described as host/parasite!

5-19 'Ecological niche' refers to the role of an organism or group of organisms in an ecosystem. It includes how the organism obtains its food, how it reproduces and its position in the food web.

5-20 Organisms that have similar requirements are in competition with each other. If they have the same niche or significant overlap they will be in strong competition. Less overlap means less competition.

5-21 Plants are the ecological group that can make their own organic materials. Through photosynthesis, plants produce simple sugars that can be used to synthesise other organic compounds. They are the organisms that form the base the vast majority of communities as they convert light energy into forms that other organisms can utilise.

5-22 Consumer types – diet and example

Consumer type	Diet	Example
Herbivore	Plant material	Cows, rabbits & koalas
Primary carnivore	Herbivores	Foxes
Secondary consumer	Primary carnivores	Wedge-tailed eagle
Decomposer	Remains/scraps of other organisms	Fungi and bacteria

5-23 Decomposers are important in all communities because they allow recycling of components of organic materials. For example, breakdown of proteins in dead animal's results in addition of nitrates to the soil. Nitrates are necessary for healthy plant growth.

5-24 Matter, over time, is recycled in ecosystems. Energy is continually being lost and there must be continuous input in the form of solar energy to keep ecosystems going.

5-25 This means the vast majority of energy is not available to the next level. Food chains are relatively short due to the rapid loss between trophic levels, i.e. 90% loss per transfer.

5-26 The arrows in food chains and food webs represent the flow of energy and matter in a community. The arrow points from the consumed to the consumer.

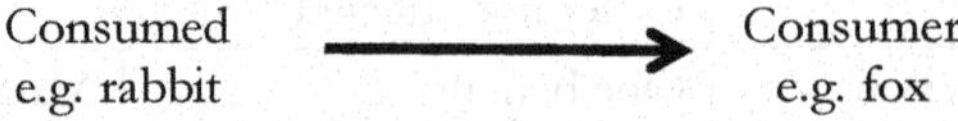

5-27 Food chain for the following for a non-vegetarian human:

5-28

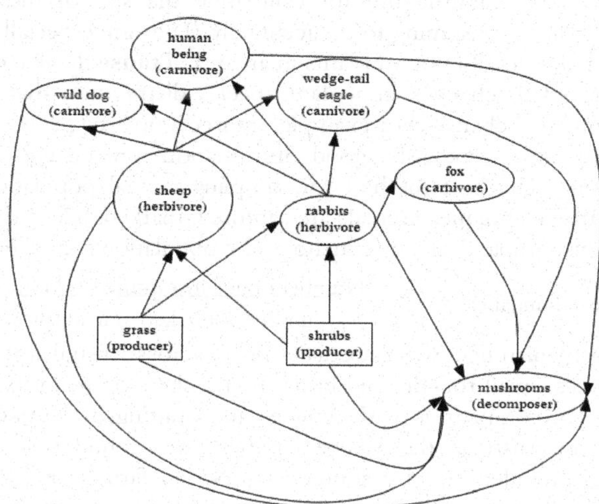

5-29 The populations of the prey of the wedge-tailed eagle (sheep, rabbits would increase. The populations of other predators (wild dogs) would increase. Grass and shrubs as they are the food source for the prey of the wedge-tailed eagle would become scarce.

5-30 Generally, the short-term consequence on an ecosystem of species competing for resources is a reduction in survival and abundance of all competing species.

5-31 Generally, the long-term consequence on an ecosystem of species competing for resources is a reduction in number of species present as usually one competitor will out compete and drive out the other species (the competitive exclusion principle).

5-32 Distribution refers to where a species is found whereas abundance refers to the how many individuals of the species are found in a particular area (density).

5-33 **A** represents a uniform distribution, **B** a clumped distribution and **C** represents a random distribution.

5-34 **B** shows a 'clumped' distribution of individuals around an important resource. If we are looking at the distribution of members of one species each will have similar needs so tend to gather around a required resource – for example: food sources, a water hole or shelter.

5-35 Sampling techniques are used when there are a large number of individuals or when the individuals are spread over a large area.

5-36 Three major methods used for sampling populations and so estimating population size are; **quadrats** (sample squares), **transect** (a line through a community) and **capture-recapture** (capture-mark-recapture) technique.

5-37 **Quadrats** are most useful for measuring population densities of organisms that do not move. Size of quadrats can be varied depending on the size of organisms being studied. A **transect** is especially useful for looking at vegetation changes across a community as the environment changes. **Capture-recapture** is useful for looking at relatively easily caught and marked populations.

5-38 **Quadrats** are of little use for estimating the size of mobile species and are relatively time consuming to do accurately. They are especially useful to gauge the number and distribution of plants in an area. **Transects** give excellent information with regard to changes in a plant species distribution that coincide with some environmental change. For example, transects up a hill, across sand dunes or rock platform give ecologists good information on changes in distribution of organisms. The best method for sampling animal populations is the '**capture-recapture**' technique. Animals are captured, marked, released and later recaptured. From this sample data, an estimate of the population can be determined.

Note: $\text{Abundance} = \dfrac{\text{Number originally captured} \times \text{Number recaptured}}{\text{number of marked individuals in recapture}}$

5-39 The Tasmanian tiger was once widespread across Australia. It ate kangaroos, other marsupials, small rodents and birds. Its numbers on mainland Australia decreased because of competition with dingoes and hunting by humans. When Europeans arrived in Australia the Tasmanian tiger was found only in Tasmania. Sheep farming provided an easy source of food, so farmers placed a bounty on their head. Finally, habitat destruction and disease led to extinction.

5-40

Evidence of changing Australian environments	**Description**
Human records - Aboriginal rock paintings	Indigenous Australians have been drawing the world around them for at least 28 000 years. The animals and plants depicted in these paintings have changed during this time, indicating changes in ecosystems.
Geological - ice core drilling	Layers indicate climates have changed and pollen and microbes indicate organisms present have changed.
- glacial erosion	Tasmanian mountain areas show evidence of glacial erosion. This indicates that some time in the past the climate was colder.
- inland limestone caves, e.g. Jenolan Caves	Limestone forms in warm shallow seas therefore these areas must once have been underwater.
Paleontological - fossils	Presence of fossils of marine organisms in outback Australia indicates that these areas were once under a sea. Sea levels have changed. Fossils indicate the types of organisms present in an area at a certain time. Fossils from Naracoorte, SA, suggest that the area was once covered with Eucalypt forests. These were replaced by open woodland trees and 10 000 years ago these were replaced by today's *Casuarina* forests.

Living organisms	
- remnant forests, e.g. *Nothofagus*	Small isolated populations are thought to be the remnants of larger more widespread populations that flourish in cool temperate environments.

5-41 Radiometric dating using carbon-14. Carbon-14 decays to form the stable isotope nitrogen-14 at a known rate. Living organisms contain the same proportion of carbon-14 as the atmosphere. When the organism dies the carbon-14 decays therefore proportion of carbon-14 remaining can be used to determine age. Carbon-14 dating is used to date charcoal used in or found with paintings or samples of once living material like wasp nests found on top of the painting. Isotopes of other elements can be used to date a microscopic layer of minerals that is deposited on the art work after it is created. The art work must be older than this layer.

5-42 Information required for radiometric dating requires
- the original proportion of the radioactive isotope used
- the half-life or rate of decay of the radioactive isotope used
- the current proportion of the radioactive isotope used.

5-43 Ice core drilling is when a cylindrical sample of ice is vertically extracted from a glacier or ice sheet. Glaciers and ice sheets form from the gradual build-up of snow. The deposition of snow is seasonal, so layers are formed. A long vertical hole is drilled through the glacier or ice sheet and a long cylindrical sample or core is extracted and the layers can be seen. The oldest layer is the deepest.

5-44 Ice cores indicate abiotic changes in climate (temperature, rainfall and composition of the atmosphere) and organisms trapped in the layers indicate biotic changes in ecosystems.

5-45 Samples are taken from the ice core and the gas is extracted by melting the sample in a vacuum. A mass spectrometer is then used to identify and quantify the molecules present.

5-46 The concentration of carbon dioxide has increased dramatically over the last 100 years. This has probably been caused by the increased use of fossil fuels since the Industrial Revolution. The burning of fossil fuels releases carbon dioxide into the atmosphere and this lead to the greenhouse effect resulting in increased temperatures.

5-47 The climate became hotter and rainfall more seasonal resulting in an arid environment.

5-48 The fossil record indicates that 150 million years ago when Australia was part of Gondwana it vegetation was dominated by conifers, cycads and ferns. About 100 million years ago flowering plants appear in the fossil record. 80-million-year-old fossil pollen from Nothofagus has been found in Australia, New Zealand and Antarctica. Mixed rainforests were common until about 25 million years ago when the climate became warmer and grasslands, woodlands and open forests began to spread. As Australia drifted northwards it became warmer and drier favouring grasslands and sclerophyll vegetation. Today Eucalypts dominate temperate areas while grasses and *Acacia* plants are the dominant plants of drier areas.

5-49 Sclerophyll vegetation refers to plants that have hard, small leaves adapted to preventing water loss. *Eucalyptus* and *Acacia* are examples of sclerophyll plants.

5-50 The ancestor of the kangaroo was a small arboreal mammal that lived in rainforests.

5-51 Food is scarce in arid environment. Hopping is a very efficient form of movement. The ability to hop allows kangaroos to travel large distances in search of food with minimum expense of energy.

5-52 The two major factors that have altered the Australian landscape since European settlement are land clearing to allow agriculture and introduction of new plant and animal species. Land clearing removed the original vegetation, which caused disruption to native communities and has often led to erosion and increased salinity. Introduction of new species such as the rabbit and fox has had a huge impact on native competitors and on suitable living space.

5-53 There is strong evidence that indigenous people changed the nature of Australian vegetation over many thousands of years through regular burning. This opened up forest areas and made areas more accessible to hunting of wildlife. Indigenous people were also thought to be responsible for removing some of the mega-fauna which existed within the last five thousand years.

5-54 Dry-land salinity has been caused by the progressive clearing of deep-rooted native vegetation and irrigation. Many Australian soils had large quantities of salt. While the water table was kept low by native vegetation, it was not a problem. With removal of the native vegetation, the water table has risen bringing salty water to the surface. In many areas, irrigation has made the problem even greater in susceptible areas.

5-55 The diseases, myxomatosis and the calicivirus, have been used to control rabbit populations. Myxomatosis is spread by mosquitoes and fleas and when initially introduced was very effective in killing rabbits. The calicivirus is spread directly from rabbit to rabbit and is most effective when population densities are high. These are examples of biological control.

5-56 Biodiversity is decreasing due to human activity, which has resulted in habitat loss and climate change.

5-57 Biodiversity is important because:
- if conditions change and biodiversity is low, there may not be a variant that can survive. This could lead to extinction not just of some species but all life on earth.
- the more diverse and ecosystem is the more resilient it is to change
- if species become extinct humans may lose potential resources, e.g. medicines.

5-58 Humans have contributed to habitat loss by clearing land for:
- agriculture
- mining
- building cities
- building roads
- logging
- damming rivers
- dredging
- waste disposal.

5-59 Habitat fragmentation occurs when a once large and continuous area is reduced to small unconnected pockets separated by very different habitats.

5-60 Habitat fragmentation results in small populations of organisms with little gene flow between them. Inbreeding occurs resulting in a reduced gene pool. If the conditions change to the detriment of one individual, then all individuals are likely to be affected leading to extinction.

5-61 Humans have impacted on ecosystems by:

- land clearing leading to habitat loss and soil erosion
- irrigation leading to increased salinity
- introducing species
- extinction of other species
- pollution
- increased carbon dioxide emissions leading to global warming, rising sea levels and ocean acidification.

5-62 Monitoring is the process of gathering information about a variable affecting a population over time. If there is a sudden change in the variable, then action can be taken to reduce the impact, for example, measuring population size of an endangered species. If the population suddenly drops then a breeding program could be used to increase numbers.

5-63 A bioindicator is a species that indicate a particular environmental condition. It is often easier to study a bioindicator than the environmental condition.

5-64 Characteristics of a good bioindicator

- sensitive to change
- react consistently to change
- representative of the other organisms in the ecosystem
- easy to observe and sample.

5-65 Scientific models are used to predict what might happen when a particular variable is changed. They are especially useful when direct experimentation is not possible. For example, it is not possible to deliberately increase greenhouse gases to see how this affects the climate. Modelling climate change allows scientist to predict what might happen in greenhouse gases increase in the atmosphere.

5-66 Scientists map the distribution and abundance of a species. Data (such as temperature, rainfall, vegetation, substrate, other organisms present) is collected from different areas within the range. This data is then used to determine the niche requirements of the species and therefore the limits to the conditions in which the species can survive. Areas where the species could survive can then be mapped. These maps can be used to determine where reserves and restoration sites could be established, search for undiscovered populations of the species and predict shifts in the species distribution due to climate change.

5-67 The GCM uses a mathematical model and computer programs to simulate the Earth's atmosphere and oceans. It is used for forecasting weather and climate change.

5-68

Process	Description
Clean up of contaminants	Contaminants are isolated and removed.
Land form reconstruction	Deep holes are filled, stabilised and reshaped.
Soil restoration	Top soil that was removed from the mine is returned and respread over the site.
Revegetation	The planting of seeds and seedlings of plants is similar to the original vegetation.
Fauna recolonisation	Construction of nesting boxes, feeding stations and control of pest species encourages animals to return.

5-69 The main causes of land degradation due to agricultural practices:
- overgrazing
- clearing of native species
- over-irrigation
- unbalanced fertiliser use.

Section I Questions

5-70 Organisms live in communities with populations of other species of organisms. Which of the following is not true?

(B) **All of an organism's needs will be met in its habitat.** (Often organisms need to move to other habitats for food or some other resource.)

5-71 In a particular area, biologists will talk about a community of organisms. Communities are made up of

(B) **populations.** (Fact)

5-72 Population size is affected by four things that will determine whether a population grows or shrinks: the interplay of birth rate (B), death rate (D), immigration rate (I) and emigration rate (E).

(C) **A population will decrease in size if D+E are greater than B + I.** (If deaths and emigration (animals leaving the population) is greater than births plus immigration (animals entering the population) then the population size will decrease.)

5-73 Exponential growth in populations does not normally occur in natural populations because

(A) **as environmental resistance increases population growth decreases.** (Environmental resistance increases as more individuals compete for the same density dependent factors – this will occur with population increase.)

5-74 Relationships between organisms in a community can be classified on the benefit or harm to those involved. Organisms where neither side is harmed include

(D) **mutualism.** (Fact – close relationships where both parties benefit.)

5-75 Using the information in the **graph**, you could conclude the following to be correct.

(B) **Species B is the predator of species A.** (Correct. Expect less of the prey species than the predator species. C may be correct, but no information is given. A & D are incorrect statements.)

5-76 Using the information in the **graph**, you could conclude the following to be correct.
(A) **There is a balance between the number of prey and predators.** (There is a balance. As prey numbers increase, despite a lag time, predator numbers increase and then decrease as prey numbers increase. For B & C there is information. The curves show a similar pattern, but they do not coincide.)
5-77 Parasite/host relationships are often very complex. Parasites have special adaptations to allow them to obtain nourishment from the host. With regard to these relationships, the most accurate statement below is
(A) **endoparasites live inside the host.** (Correct statement. Exoparasites live on the skin of a host. While parasites generally do not kill their hosts, they do harm them and often cause significant health problems.)
5-78 Mutualism and commensalism are similar in that both members of the relationship
(C) **neither are harmed.** (In symbiosis, both benefit and neither are harmed. In commensalism neither are harmed although only one benefits.)
5-79 Organisms in a community can normally be classified into three categories
(A) **producers, consumers & decomposers.** (Fact.) The other alternatives contain repeats. For example, in B the terms producers and autotrophs refer to the same groups.)
5-80 Heterotrophs are classified on what they eat and where they are in a food chain. The most complete correct statement is
(C) **decomposers break down dead remains.** (Correct – herbivores eat plant material, carnivores only eat meat, and omnivores eat both plant and animal material.)
5-81 Heterotrophs require digestive systems because
(B) **the molecules in the food they eat are too big to pass into the body.** (Heterotrophs can build organic molecules, but they must consume the basic building blocks of these molecules.)
5-82 In food chains and food webs, the arrows are best described as presenting
(B) **the movement of matter and energy from one trophic level to the next highest level.** (Fact. A is partly correct but not the best answer. C & D describe energy or matter moving the wrong way.)
5-83 With regards to energy movement through a community, scientists often refer to the '10% rule'. The rule refers to
(B) **the idea that roughly 10% of the energy harnessed at one trophic level is passed onto the next level.** (Fact)
5-84 In a community, energy is passed from one trophic level to the next. Most communities are governed by the 10% rule – that is only 10% of the energy at one level is available to be passed on to the next level. As a consequence of this rule
(B) **a large biomass of producers is usually required to support the rest of the community.** (Energy is lost as it is passed from one level to the next. Roughly 10% is passed on so usually need a large biomass of producers to support the rest of the community and food chains are usually short.)
5-85 An Australian rural area has the following organisms present: grasses, rabbits, wedge-tailed eagles, mushrooms, rabbits, sheep and foxes. A likely food chain is
(C) **grass, rabbit, fox, mushroom.** (B & C are possible answers. C is better as it only contains organisms mentioned in the question stem.)
5-86 The organisms capable of photosynthesis in the food web are:
(A) **zooxanthellae.** (Zooxanthellae are symbiotic algae that live in coral tissue and photosynthesise.)

5-87 First order consumers in the food web are:
(B) **zooplankton.** (First order consumers feed directly on producers. Zooplankton in feeding on a producer, phytoplankton, is a 1st order consumer.)
5-88 The longest food chain in the web contains
(B) **7 organisms.** (Sea cucumber feeds on a chain that includes six other organisms.)
5-89 Matter and energy flow through ecosystems. A major difference between energy and matter in ecosystems is
(D) **matter is recycled whereas as an input of energy is required to replace energy lost from an ecosystem.** (Main point is that matter is recycled and energy is not)
5-90 Scientists in studying different environments need to gather data to measure physical conditions. Which of the following is **incorrect**?
(D) **wind speed can be measured with a hygrometer.** (Wind speed is measured with an anemometer)
5-91 Different methods are appropriate for determining the numbers and distributions of different species. Quadrats are useful for
(B) **distribution of a particular plant species in an area.** (A quadrat is a marked area, usually a square, and gives useful information on distribution.)
5-92 Transects are very useful for providing information about communities for all but one of the following
(D) **determining the change in population density of plants in a pasture.** (Transects are most useful in determining changes to species present over a range rather than giving actual numbers or densities.)
5-93 The capture-recapture method relies on which of the following factors:
(C) **predation of individuals is not changed by marking technique.** (The marking must not alter the individual's chances of survival. All individuals are assumed to have the same chance of surviving. Individuals may die – marked or unmarked should die at the same rate.)
5-94 The best technique to look at changes in distribution of plant life across a sand dune would be
(D) **a transect.** (Provides a line survey through a community)
5-95 In using the 'capture–recapture' method one of the following is not an assumption
(A) **marking makes the animal more susceptible to predation.** (This alternative is not an assumption. For the method to work captured animals, after marking, cannot be at a disadvantage. If they were the method would consistently underestimate populations.)
5-96 The extinction of the Tasmanian tiger was due to
(D) **all of the above.** (A bounty was place on Tasmanian tigers, farming destroyed their habitat and disease killed many.)
5-97 It is difficult to date rock paintings because
(D) little organic material remains. (Carbon-14 is used to date organic remains up to 50 000 years old. Little organic material was ever present the paints used.)
5-98 This suggests the paintings are
(C) **older than 17 000 years old.** (The rock was painted before the wasps laid their nests therefore the paintings must be older than the wasps nest.
5-99 The layers seen in ice cores are formed from
(B) **compressed snow.** (Compressed sediments form sedimentary rock. Ice cores are taken from glaciers or ice sheets. These are formed from seasonal deposition of snow.)

5-100 Ice cores indicate
(C) **both biotic and abiotic factors present at the time the layers were formed.** (The thickness of the layers suggests past temperatures; trapped gases indicate the composition of the atmosphere and the presence of pollen and microbes indicate what organisms were present.)
5-101 As Australia broke away from Gondwana and moved north, the climate became
(A) **drier and hotter.** (Australia moved north towards the equator and therefore became hotter. Today Australia is described as arid. This means that it has a dry climate.)
5-102 Which of the following adaptations is **not** an adaptation for living in a hot, dry climate?
(A) **ability to sweat.** (Sweating results in water loss and this is a disadvantage in a dry environment.)
5-103 Decreased biodiversity
(B) **increases the chances of extinction.** (If all organisms are similar and one is affected to its detriment, all individuals will suffer.)
5-104 Organisms that are good bioindicators
(C) **are sensitive to change and are easy to observe and sample.** (Bioindicators must be able to react to change in a consistent way that is easy to observe.)
5-105 Scientific modelling allows scientists to
(A) **predict what might happen to biodiversity in the future**
(Modelling allows scientist to predict what may happen if a variable is altered without performing the experiment. How accurate the prediction is dependent on the data used in the model and the quality of the computer programs used.)

Section II Questions

5-106

(a) populations and communities – a population consists of organisms from one species living in an area. A community refers to all of the different species – that is different populations living in an area. (Example: A grassland community might contain rabbit populations, kangaroo populations and various different plant species.)

(b) biotic factors and abiotic factors – biotic factors are related to organisms and/or their products and the effect they have on other organisms. Abiotic factors are those that are related to the non-living physical environment. (For a koala in a gum tree, biotic factors include competition from other organisms including koalas, quantity of leaves available and any diseases that might affect koalas. Abiotic factors include soil in which their trees grow, air quality, temperature and in general climatic conditions.)

5-107

	Relationship	Reason
tapeworm in pigs	parasite/host	tapeworm benefits, pig is harmed
sheep and kangaroos in a paddock	inter-specific competitors	interaction harms each other – reducing food resource available.
lichen – consisting of a fungus and algal cells	symbiosis or mutualism	both organisms benefit. The fungus provides shelter & the algal cells produce organic compounds.
sea anemone catching small fish	predator/prey	anemone obtains food and fish dies.
mosquito on you!	parasite/host	mosquito receives nutrients (benefit) & human gets a nasty bite and possible infections (harm).
rabbits in a paddock	intra-specific competition	rabbits competing with each other for food – both harmed.

5-108

(a) (i) The term 'producer' refers to the fact that they can manufacture their own organic material – produce usually via photosynthesis. Autotroph refers to the concept that the organism is the source of its organic material rather than relying on other organisms. In practice the two terms are used interchangeably.

(ii) The term 'consumer' refers to the organism eating or consuming other organisms to procure their organic material needs. Heterotroph refers to the concept that organic material comes from a source other than from itself. In practice the two terms are used interchangeably.

(b) Complete the following table adding the most appropriate description of the relationship between the two organisms in the right-hand column.

Organisms	Relationship
Two sheep in a paddock	intra-specific competition
A sheep and a rabbit in a paddock	inter-specific competition
Koala and a gum tree	predator/prey
Fungi and algal components of lichen	mutualism or symbiosis
Flea and a dog	parasite and host
Mould and some old bread	decomposer on dead organic matter

(c) clown fish and its anemone.
shark and a cleaner wrasse.

(d) Probably not! It is thought that clownfish might offer some protection against fish that graze the anemones. The wrasse gets a food supply. It may be of benefit for the shark to have external parasites removed. The wrasse would probably slow the shark slightly – a harm.

5-109

(a) – collapse of herbivore populations and subsequently higher consumers
– replacement of producer by another producer species

(b) – numbers of rabbits and feral cats increases.
– impact on producers, i.e. over eating of grass by rabbits

(c) – other herbivores have greater food supply
– less consumption of producer species. (Historically, after the removal of rabbits, pastures recovered dramatically.)

(d) – removal of native forests
– hunting of native animals – sometimes to extinction.
– introduction of new plant and animal species
– irrigation leading to salting.

(e) Any solutions must be global solutions. The atmosphere does not know national boundaries; therefore, all nations must undertake to reduce production of greenhouse gases if we are to have an impact. Nevertheless, the developed world, as early and dominant producers of greenhouse gases, has a responsibility to lead and set an example.

5-110

(a) Abundance = 50/5 = 10 *Salicornia* plants per m^2

(b) 10 × 80 = 800 *Salicornia* plants

(c) The students' results are not very valid as only 5 quadrats were used. Furthermore, the number of plants in each quadrat showed great variation. Sampling more quadrats would take into account clumping of plants thus giving a more accurate estimation.

(d) No. Quadrats are not accurate for populations that move. Individuals may move into and out of quadrats therefore may not be counted or may be counted more than once.

5-111

(a)

$$\text{Abundance} = \frac{\text{Number originally captured} \times \text{Number recaptured}}{\text{number of marked individuals in recapture}}$$

$$= \frac{20 \times 10}{4}$$

$$= 50 \text{ pigmy-possums}$$

(b) The results for this sampling are not very valid as only 20 possums were originally caught and tagged and only 10 were recaptured. More possums would need to be tagged and the process repeated many times for the results to be statistically valid.

5-112

(a) At first, when Australia was part of Gondwana, the dominate vegetation was forests of conifers, cycads and ferns. About 100 million years ago, flowering plants appeared and the dominant vegetation was that of mixed forests.

(b) After Australia separated from Gondwana, the rainforests began to contract. They were replaced with grasslands, woodlands and open forests. Today, Eucalypts dominate temperate areas while grasses and Acacia plants are the dominant plants of drier areas.

(c) Climate change. The climate has become hotter and drier. Eucalypts, grasses and Acacia are well adapted to dry, hot environments.

5-113

(a) transect

(b) Change in rainfall – lowest in desert regions, gradually rising over the divide and highest near the coast and rainforest.
Change in temperature – highest in the desert region, gradually dropping over the divide but rising again on the coastal lowlands.

(c) Rainforest – dense canopy shades the ground, little direct view of the sky, ground moist and sheltered from wind.
Woodland – light canopy, sky little blocked by canopy, light reaches the ground, ground dry, and exposed to wind.

(d) Much of the sediment from the rivers has ended up coating the coastal reef. Healthy coral requires light to allow photosynthesis in symbiotic bacteria. The sediment reduces the light that reaches the coral and other photosynthetic organisms.

(e) Clearing that left strips of the original vegetation along rivers and creeks would have reduced the runoff of sediments into water courses and so reduced the sediment washed out onto the reef.

(f) Mangroves trap sediments and reduce erosion and thus improve water quality. The mangroves provide food and shelter for the juvenile of many fish species. Altering these areas reduces the suitable places for the fish to reproduce and develop thus reducing fish populations.

Chapter 7
Sample Examination Paper

This chapter contains a sample examination paper. When you sit your school's examination you should note the following instructions.

- The paper consists of two sections – Section I and Section II.
- Answer **all** questions from Section I. Section I is worth 20 marks.
- You should spend approximately 35 minutes answering Section I.
- In Section I you choose the response that is **correct** or **best answers the question**.
- In Section I a correct answer is worth 1 mark, an incorrect answer is worth no marks. No mark will be given if more than one answer is shown for any one question. Marks will not be deducted for incorrect answers. You should **attempt every question**.
- Answer all questions from Section II. Section II is worth 80 marks.
- The paper is worth 100 marks. Allocate time appropriate to the number of marks.
- Check that you answer each page.
- You have 5 minutes reading time followed by 180 minutes to write your answers.
- Write in legible English using an ink or ballpoint pen. (Examiners find it hard to read and mark messy answers – especially when written in pencil.)

Section I Questions – 20 marks

Attempt Questions 1–20.

Allow 35 minutes for this part.

Use the following information to answer Questions 1 and 2.

The following diagram represents the field of view of a light microscope fitted with a 10x eyepiece and a 40x objective lens.

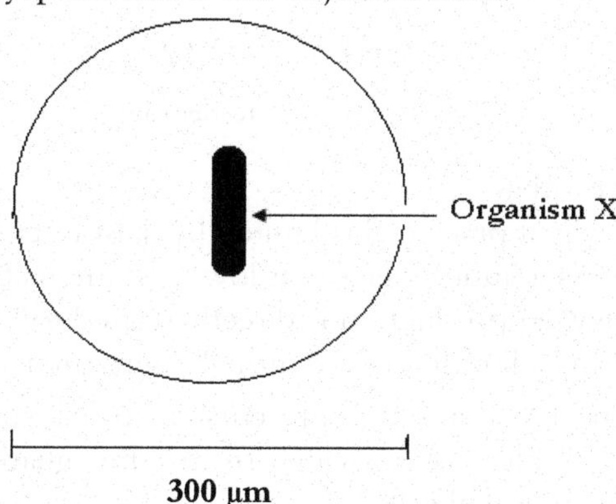

1 The length of organism X is

(A) 30 μm.
(B) 100 μm.
(C) 200 μm.
(D) 230 μm.

2 Organism X has been magnified by

(A) 10x.
(B) 40x.
(C) 200x.
(D) 400x.

3 In photosynthesis, producers convert light energy to chemical energy. One of the following is not directly necessary for photosynthesis.

(A) carbon dioxide
(B) water
(C) chlorophyll
(D) oxygen

4 Outputs of aerobic respiration are

(A) oxygen, water and light.
(B) glucose and carbon dioxide.
(C) carbon dioxide, water and ATP.
(D) carbon dioxide and water.

5 Scientists carried out an experiment to investigate the effect of light intensity on the rate of photosynthesis. The following results were obtained.

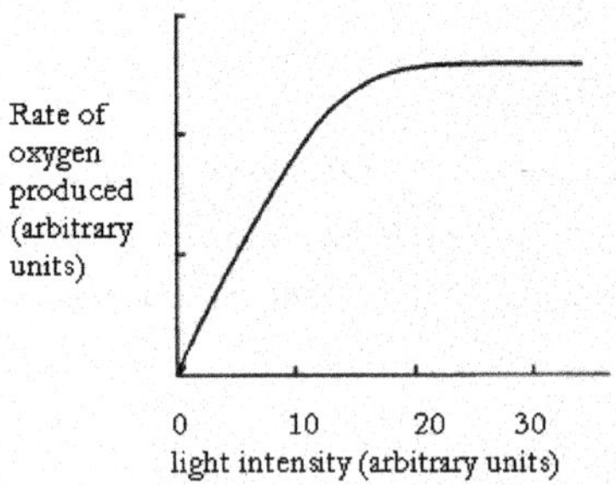

These results suggest that

(A) the rate of oxygen produced is limited by light intensity.
(B) light intensity is a limiting factor at low light intensities.
(C) the rate of oxygen production is directly related to light intensity.
(D) light intensity is a limiting factor at high light intensities.

6 The diagram that follows represents a stoma surrounded by two guard cells. Water is moving from the guard cells into the surrounding epidermal cells.

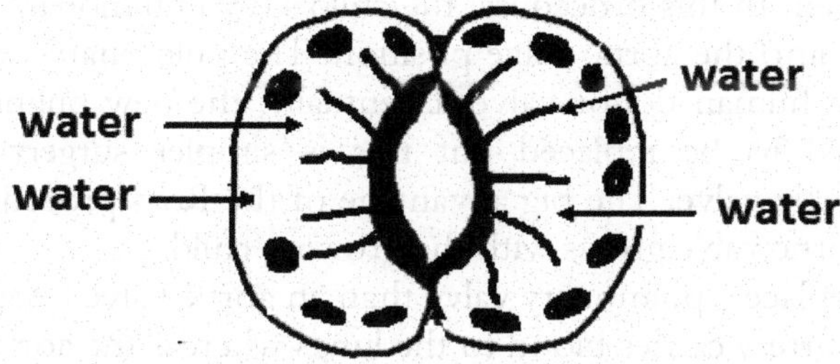

As a result,

(A) the size of the stoma will increase.
(B) the size of the stoma will decrease.
(C) more water will be lost from the plant.
(D) photosynthesis will cease in the guard cells.

Use the following information to answer Questions 7 and 8.

The graph that follows shows the percentage saturation of haemoglobin with oxygen at different oxygen partial pressures.

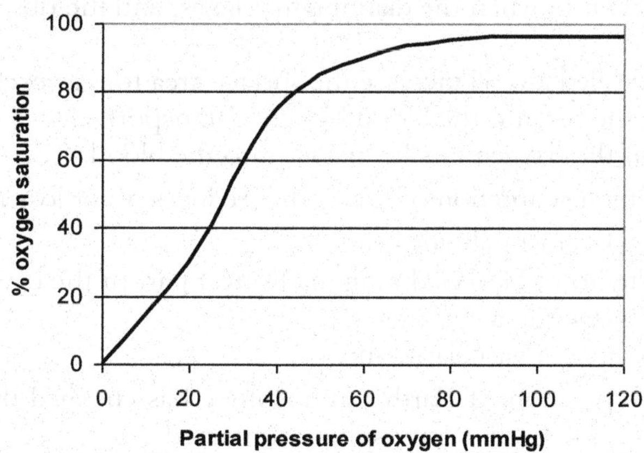

7 Blood leaving the lungs has a partial pressure of oxygen of 100 mmHg. The percentage oxygen saturation of haemoglobin is

(A) 25%.
(B) 50%.
(C) 95%.
(D) 100%.

8 During exercise, the partial pressure of oxygen in the tissue drops to 10 mmHg. The percentage of oxygen made available to the tissue is

(A) 40%.
(B) 60%.
(C) 80%.
(D) 100%.

9 The Ross procedure is used to replace aortic valves in the damaged hearts of children. In this procedure, the child's own pulmonary valve is transplanted into the aortic valve position. The pulmonary valve is replaced with a human donor valve. Eventually, the new pulmonary valve may need to be replaced but this is simpler surgery than replacing an aortic valve. The big advantage of the Ross procedure is that the new aortic valve grows with the growing child.
It is easier to replace a pulmonary valve than an aortic valve because

(A) the pulmonary artery carries blood to the lungs whereas the aorta carries blood under huge pressure to the whole body.
(B) the aorta carries oxygenated blood whereas the pulmonary artery carries deoxygenated blood.
(C) the pulmonary valve directs blood flow between the right atrium and ventricle whereas the aortic valve directs blood between the left ventricle and the rest of the body.
(D) the aorta directs blood to the lungs and the pulmonary artery directs blood to the heart.

10 In the lungs

(A) air is drawn in when the diaphragm relaxes, and the ribs move up and out.
(B) air sacs called alveoli increase the surface area for gas exchange.
(C) oxygen moves into the blood by active transport.
(D) 100% of the oxygen inhaled passes into the blood.

11 A structural adaptation of leaves that reduces water loss is

(A) a double layer of palisade cells.
(B) the ability to roll leaves during the hottest part of the day.
(C) a large surface area.
(D) the presence of vascular tissue.

12 The composition of Earth's atmosphere has changed throughout its history. Early Earth's atmosphere contained

(A) low levels of carbon dioxide.
(B) high levels of ozone.
(C) little free oxygen.
(D) high levels of ammonia.

13 In order from oldest to most recent, the major stages in the evolution of living things is

(A) anaerobic eukaryotes, photosynthetic eukaryotes, aerobic eukaryotes, prokaryotes.
(B) aerobic prokaryotes, anaerobic prokaryotes, photosynthetic prokaryotes, eukaryotes.
(C) photosynthetic prokaryotes, aerobic prokaryotes, anaerobic prokaryotes, eukaryotes.
(D) anaerobic prokaryotes, photosynthetic prokaryotes, aerobic prokaryotes, eukaryotes.

14 ^{14}C has a half-life of approximately 6000 years. If 25% of the original ^{14}C remains in a fossil, the fossil is

(A) 6000 years old.
(B) 9000 years old.
(C) 12 000 years old.
(D) 15 000 years old.

15 Approximately 70% of living marsupial species are found in Australia. This is an example of

(A) convergent evolution.
(B) adaptive radiation.
(C) selective radiation.
(D) punctuated equilibrium.

16 Fossils in sedimentary rocks and ice cores are used to study changes in ecosystems over time. Which of the following statements is correct?

(A) The oldest layers in both sedimentary rocks and ice cores are the layers closest to the surface.
(B) The oldest layers in both sedimentary rocks and ice cores are the deepest layers.
(C) The oldest layers in sedimentary rocks are the deepest but in ice cores the oldest layers are those closest to the surface.
(D) The oldest layers in ice cores are the deepest but in sedimentary rocks the oldest layers are those closest to the surface.

17 Mass spectrometer analysis of gas trapped in ice cores from Antarctica indicates

(A) temperatures and carbon dioxide concentration in the atmosphere have increased.
(B) temperatures and carbon dioxide concentration in the atmosphere have decreased.
(C) temperatures have decreased but carbon dioxide concentration in the atmosphere has increased.
(D) temperatures have increased but carbon dioxide concentration in the atmosphere has decreased.

18 Scientific modelling allows scientists to

(A) investigate how changes may affect species without actually making those changes.
(B) be certain of how climate change will affect other species.
(C) reduce experimental error.
(D) obtain accurate results.

19 The clearing of land by early European settlers in Australia resulted in

(A) an increase in biodiversity and extinction of many species.
(B) an increase in biodiversity but did not affect extinction rates.
(C) a decrease in biodiversity and extinction of many species.
(D) a decrease in biodiversity but did not affect extinction rates.

20 Which of the following practices could not be used to restore a damaged ecosystem?

(A) planting of seedlings endemic to the area
(B) introduction of new species
(C) returning the original top soil to an area
(D) removing contaminants from the area

Section II Questions – 80 marks

Attempt Questions 21–31.

Allow 2 hours 25 minutes for this part.

Question 21

The diagram below is of a cell seen under an electron microscope.

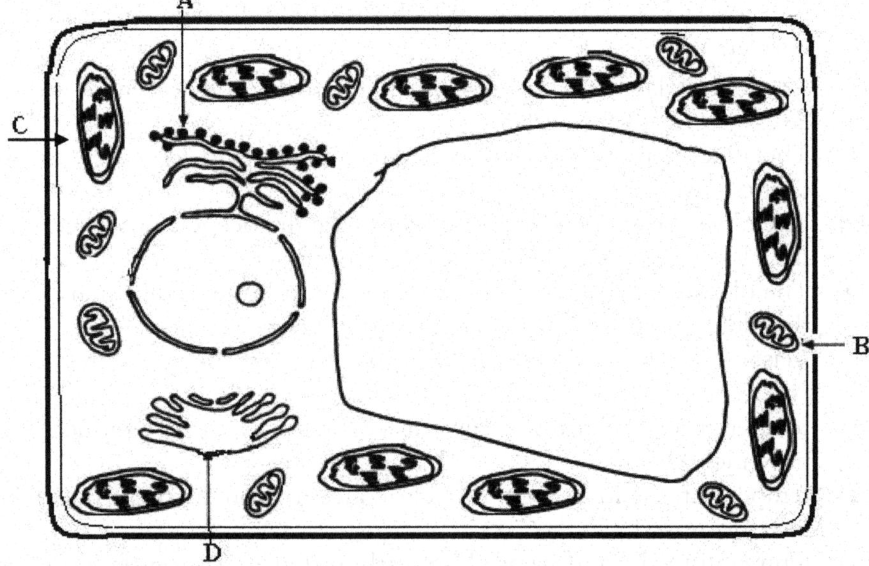

(a) Is the cell an animal or plant cell? Explain your answer. [2 marks]

(b) Complete the following table for structures labelled in the diagram.

	Structure	Function
A	ribosome	
B		major site of ATP production
C	chloroplast	

[3 marks]

(c) Which of the above structures would be most clearly seen with a light microscope? Explain your answer. [1 mark]

(d) Often stains are used to improve contrast and to locate specific molecules. Name one disadvantage of using a stain. [1 mark]

[Total 7 marks]

Question 22

The diagram below is of a cell membrane.

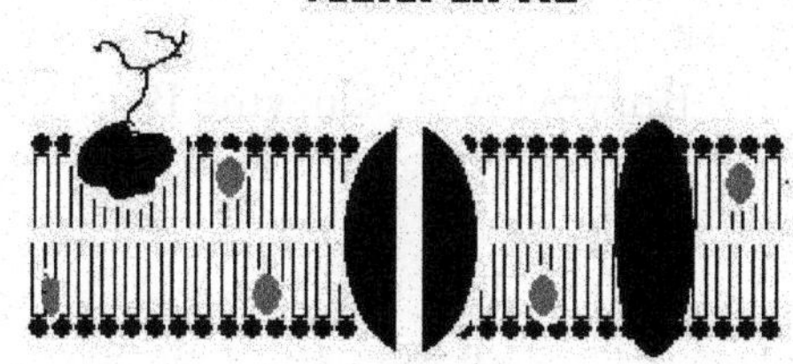

(a) What are the two major compounds that make up a cell membrane and what is the function of each in the membrane? [2 marks]

A scientist performed the following experiment. Rhubarb was cut into 30 pieces of similar size. Ten pieces were placed into each of three Petri dishes and then covered with sucrose solution. The concentration of sucrose in the first dish was 1.5 M, the second 0.3 M and the third 0.0 M. The Petri dishes were covered and placed next to each other in a cupboard. The next day the rhubarb pieces were examined under a microscope.

(b) What is the independent variable in this experiment? [1 mark]

(c) Why is it important that the Petri dishes were placed next to each other in a cupboard? [1 mark]

(d) Match the following diagrams with the cells you would expect to see in each solution. [1 mark]

Cell A

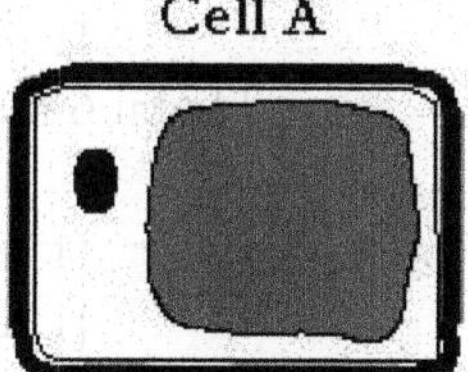

Cell B

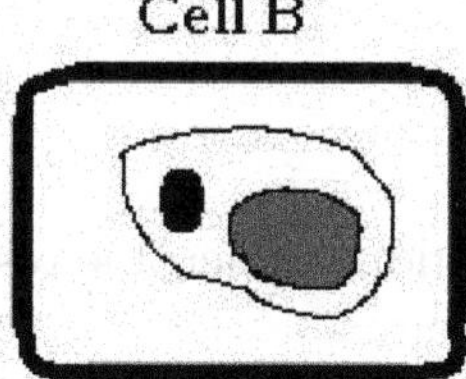

Cell C

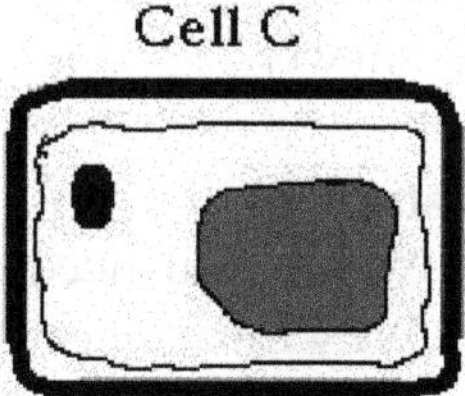

(e) Explain what happened to cell B. [2 marks]

[Total 7 marks]

Question 23

CSIRO has developed an enzyme-based product that can rapidly breakdown pesticide residues in soil and water.

(a) What is an enzyme? [1 mark]

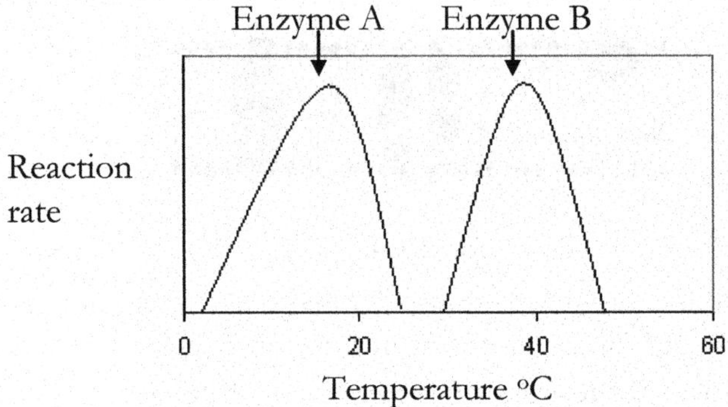

(b) Which enzyme, A or B, is most likely to be the enzyme developed by CSIRO? Explain your choice. [2 marks]

(c) A scientist heated enzyme B to 50°C for 10 minutes and then let it cool down to 38°C. Would this enzyme still function? Explain your answer. [2 marks]

Enzymes are sometimes used to breakdown oil spills at sea. Usually they are used as a final step after more conventional clean-up techniques have been used.

(d) Could enzyme A also be used to clean up oil spills? Explain your answer. [1 mark]

[Total 6 marks]

Question 24

(a) Write a balanced equation for aerobic respiration. [1 mark]

(b) Why must all cells respire? [1 mark]

(c) What are the inputs and outputs in photosynthesis? [1 mark]

Use the following diagram to answer parts ***(d)*** *and* ***(e)****.*

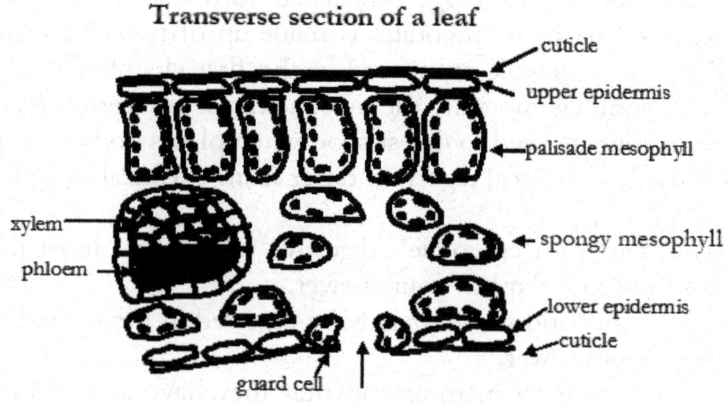

(d) Would you expect to find more or fewer chloroplasts in palisade cells than in spongy mesophyll cells? Explain your answer. [1 mark]

(e) Describe two ways guard cells are important to vascular plants. [2 marks]

Scientists carried out an experiment to investigate the effect of light intensity on the rate of photosynthesis.

The following results were obtained.

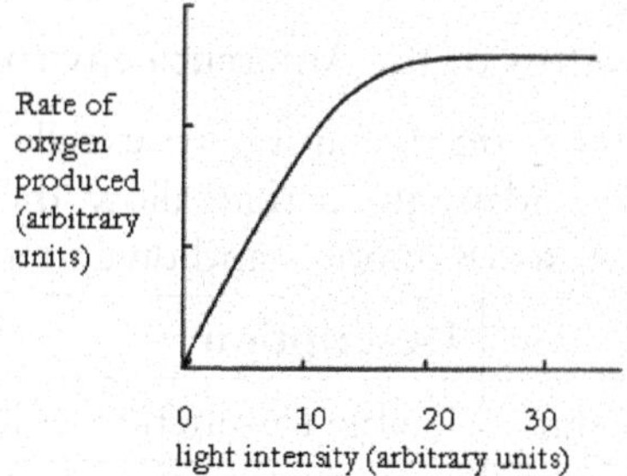

(f) Why can 'rate of oxygen produced' be used as a measure of photosynthetic rate? [1 mark]

(g) Is rate of oxygen production a measure of gross or net photosynthetic rate? Explain your answer. [2 marks]

(h) Suggest a reason why the rate of oxygen produced levels off at around a light intensity of 20 arbitrary units? [1 mark]

(i) What element is added to the carbohydrate that plants produce in photosynthesis to make protein? [1 mark]

[Total 11 marks]

Question 25

Crocodiles have powerful jaws and very sharp teeth. They eat small animals by swallowing them whole. Large animals are torn apart before swallowing. The muscular stomach of crocodiles is made up of two chambers. Stones swallowed by the crocodile are stored in the first chamber. The digestive juices in the second chamber are the most acidic of any vertebrate.

(a) Why do animals need to digest food when plants do not? [2 marks]

(b) How does mechanical digestion differ from chemical digestion? [2 marks]

(c) In what part of a crocodile's digestive tract would most mechanical digestion occur? Explain your answer. [2 marks]

(d) What type of food would mainly be digested in the second chamber? Explain your answer. [2 marks]

Crocodiles are similar to mammals in that they have a closed circulatory system. Also, they are one of the few reptiles that have a four-chambered heart like birds and mammals.

(e) In closed circulatory systems, where does the exchange of materials between the blood and the cells occur? [1 mark]

(f) What are two advantages of a four-chambered heart? [2 marks]

Question 26

The emperor penguin (*Aptenodytes forsteri*) lives in Antarctica. It is the tallest and heaviest penguin and like all penguins is flightless, has a streamlined body and flattened wings that act as flippers.

(a) List two abiotic factors of the Antarctica environment. [1 mark]

The emperor penguin feeds on fish and is an excellent swimmer. The table below lists some of the adaptations that allow the emperor penguin to remain submerged for up to 18 minutes and dive to a depth of 535 m.

	Adaptation	**Description**
1	Unusual haemoglobin	Is able to function at low oxygen levels
2	Solid bones	Reduces damage due to change in pressure during a dive

(b) Classify each adaptation as structural, physiological or behavioural. [2 marks]

(c) Describe how the penguin flipper may have evolved by natural selection. [3 marks]

(d) Penguins, dolphins and sharks have a streamlined shape. Is this an example of divergent or convergent evolution? Justify your answer. [2 marks]

[Total 8 marks]

Question 27

Cytochrome c is a respiratory enzyme found in the mitochondria of cells. Comparison of cytochrome c amino acid differences is sometimes used to determine evolutional relationships. The table below shows the number of amino acid differences between humans and a number of organisms.

Species	Number of differences from human cytochrome c
Chimpanzee	0
Fruit fly	29
Horse	12
Pigeon	12
Rattlesnake	14
Bread mould	48
Rhesus monkey	1

(a) On the basis of cytochrome c comparisons, which organisms are:

(i) most closely related to humans? [1 mark]

(ii) least closely related to humans? [1 mark]

(b) Explain why cytochrome c comparisons can be used to determine evolutionary relationships. [3 marks]

(c) Both horse and pigeon cytochrome c vary from human cytochrome c by 12. Does this mean that horses and pigeons are closely related? Explain your answer. [1 mark]

[Total 6 marks]

Question 28

Complete the following table, adding the most appropriate description of the relationship between the two organisms in the right-hand column.

Organisms	Relationship
Two goats in a paddock	
Fungi and algal components of a lichen	
Wedge-tailed eagle and rabbit	

[Total 3 marks]

Question 29

Examine the food web for a soil community that follows. Use the information in the web to answer the questions that below.

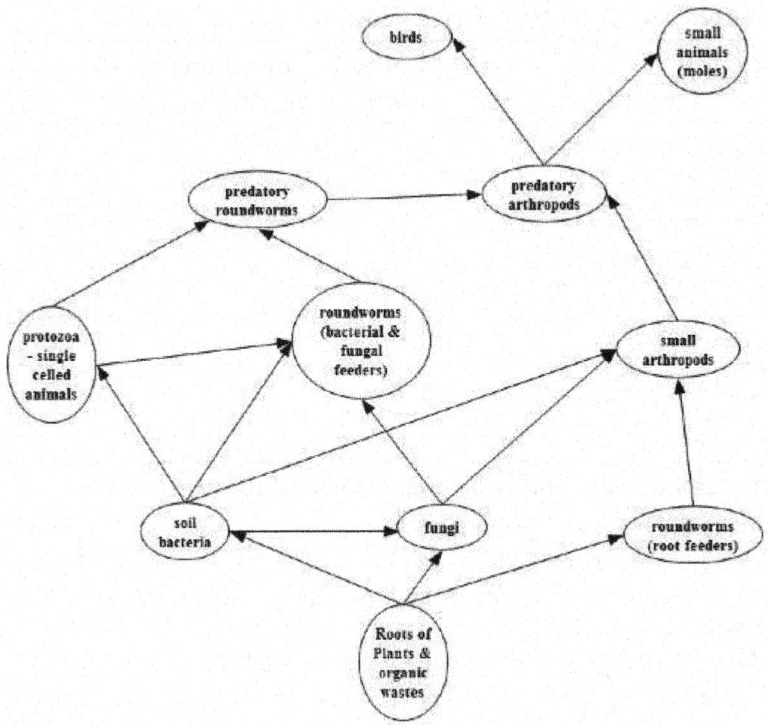

(a) Name the producer(s) in this food web. [1 mark]

(b) What is the source of energy for the organism(s) named in (a)? [1 mark]

(c) What organisms are in the second trophic level? [1 mark]

(d) In terms of biomass, which trophic level would you expect to be largest and which trophic level the smallest? [1 mark]

(e) Explain why you have given this answer. [1 mark]

(f) Which organisms are present but could have been included again with new links in this food web and what in general is their role in a community? [2 marks]

[Total 7 marks]

Question 30

To study the changes after a drought, a group of biologists carried out a field survey of the density and distribution of a small flowering plant, *Bellis perennis* (common European daisy). The ecologists used 1 m square quadrats to sample the area that was being examined.

Below is a diagram of the 100 square metre area (10 m by 10 m) and the distribution of the plant; the numbers in 10 sample quadrats are the plants counted.

<table>
<tr><td>4</td><td></td><td>7</td><td></td><td></td><td></td><td></td><td></td><td></td><td></td></tr>
<tr><td></td><td></td><td></td><td></td><td></td><td></td><td>11</td><td></td><td></td><td></td></tr>
<tr><td></td><td>9</td><td></td><td></td><td></td><td></td><td></td><td></td><td></td><td></td></tr>
<tr><td></td><td></td><td></td><td>4</td><td></td><td></td><td></td><td></td><td></td><td></td></tr>
<tr><td></td><td></td><td></td><td></td><td></td><td></td><td></td><td></td><td>9</td><td></td></tr>
<tr><td></td><td></td><td>7</td><td></td><td></td><td></td><td></td><td></td><td></td><td></td></tr>
<tr><td></td><td></td><td></td><td></td><td></td><td></td><td></td><td></td><td></td><td></td></tr>
<tr><td></td><td>9</td><td></td><td></td><td></td><td></td><td></td><td></td><td>10</td><td></td></tr>
<tr><td></td><td></td><td></td><td></td><td>7</td><td></td><td></td><td></td><td></td><td></td></tr>
<tr><td></td><td></td><td></td><td></td><td></td><td></td><td></td><td></td><td></td><td></td></tr>
</table>

(a) What is the estimated population of this plant in the whole area? [2 marks]

(b) What is the average population density in the whole area? Show your working. [2 marks]

(c) Name an abiotic factor that might influence the distribution of the daisy. [1 mark]

The 'mark and recapture' method is used to determine the population of mobile animals. To estimate the population of an area, the following formula is used.

$$\text{Total Population} = \frac{\text{No. of animals in 1st sample (all marked)} \times \text{Total no. of animals in 2nd sample}}{\text{Number of marked animals in the 2nd sample (recaptured)}}$$

A scientist is trying to estimate the population of rabbits in an island that is to be exposed to the calicivirus. Samples were taken 6 months before the release of the virus and 6 months after the release of the virus. The figures are shown in the table that follows.

	6 months before	6 months after
size of 1st sample	195	55
size of 2nd sample – week later	250	65
Number marked in 2nd sample	50	15

(d) Using the formula above, calculate the figure for the two sampling periods before and after the virus release. [2 marks]

(e) Approximately how effective has the virus been at reducing the rabbit population? Justify your answer. [2 marks]

[9 marks]

Question 31

The southern beech (*Nothofagus*) is found in the temperate areas of South America, Australia, New Zealand, New Guinea and New Caledonia. It is not found in Africa.

(a) Given your knowledge of Gondwana, explain the present-day distribution of the southern beech. [1 mark]

(b) Fossil pollen has been found in Antarctica. What does this indicate about the past climate of Antarctica? [1 mark]

(c) What conditions are required for fossilisation? [2 marks]

(d) Describe one other piece of evidence supporting the existence of Gondwana. [1 mark]

[Total 5 marks]

Chapter 8
Suggested Answers: Sample Examination Paper

Section I Questions – Suggested Answers

The best alternative for each question is given. In brackets after the correct alternative are some comments as to why the other alternatives are incorrect.

7-1

The length of organism X is
(B) 100 μm. (Organism X is about the length of 1/3 of the field of view and the field of view is 300 μm.)

7-2

Organism X has been magnified by
(D) 400x. (Magnification equals 10 x 40 = 400x.)

7-3

In photosynthesis, producers convert light energy to chemical energy. One of the following is not directly necessary for photosynthesis
(D) oxygen (Oxygen is a product of photosynthesis – not a reactant)

7-4

Outputs of aerobic respiration are
(C) carbon dioxide, water and ATP. (Oxygen and glucose are inputs of aerobic respiration.)

7-5

These results suggest that
(B) light intensity is a limiting factor at low light intensities. (Up until approximately 15 arbitrary units of light intensity the rate of oxygen produced is directly proportional to light intensity suggesting that light intensity is the limiting factor. After 15 arbitrary units, increasing light intensity does not increase the rate of oxygen production therefore some other factor must be limiting oxygen production.)

7-6

As a result
(B) the size of the stoma will decrease. (As water is lost from guard cells the stoma closes.)

7-7

The percentage oxygen saturation of haemoglobin is
(C) 95%. (Move along the X axis to 100, draw a line up until the curve is intersected, then draw a line over to the Y axis.)

7-8

The percentage of oxygen made available to the tissue is
(C) 80%. (In the lungs the blood is 95% saturated. In the tissue the blood is 15% saturated therefore 95 – 15 = 80% available to the tissue.)

7-9

It is easier to replace a pulmonary valve than an aortic valve because

(A) the pulmonary artery carries blood to the lungs whereas the aorta carries blood under huge pressure to the whole body. (The pulmonary valve directs blood from the right ventricle to the lungs. Whether the blood is oxygenated or deoxygenated is not relevant but the direction of flow to the body is.)

7-10

In the lungs

(B) air sacs called alveoli increase the surface area for gas exchange. (Air is drawn in when the diaphragm contracts, oxygen diffuses into the blood and 100% of the inhaled oxygen does not pass into the blood.)

7-11

A structural adaptation of leaves that reduces water loss is

(B) the ability to roll leaves during the hottest part of the day. (A) and (D) have no direct influence on water loss. (C) is incorrect as a large SA tends to increase water loss.

7-12

The composition of Earth's atmosphere has changed throughout its history. Early Earth's atmosphere contained

(C) little free oxygen. (There was no ozone layer, low levels of ammonia and high levels of carbon dioxide in early Earth's atmosphere.)

7-13

In order from oldest to most recent, the major stages in the evolution of living things is

(D) anaerobic prokaryotes, photosynthetic prokaryotes, aerobic prokaryotes, eukaryotes. (Prokaryotes occurred before eukaryotes. The atmosphere of early earth had little oxygen therefore anaerobic prokaryotes were first. Photosynthetic prokaryotes added oxygen to the atmosphere which could then be used by aerobic prokaryotes.)

7-14

^{14}C has a half-life of approximately 6000 years. If 25% of the original ^{14}C remains in a fossil, the fossil is

(C) 12 000 years old. (After 6000 years or 1 half-life, 50% would remain. After another 6000 years half of 50% or 25% would remain.)

7-15

Approximately 70% of living marsupial species are found in Australia. This is an example of:

(B) adaptive radiation. (The marsupials have a common ancestor that has given rise to many new species occupying different niches.)

7-16

Fossils in sedimentary rocks and ice cores are used to study changes in ecosystems over time. Which of the following statements is correct?

(B) The oldest layers in both sedimentary rocks and ice cores are the deepest layers. (The oldest layers in both sedimentary rocks and ice cores are the layers that form first and therefore are the deepest.

7-17

Mass spectrometer analysis of gas trapped in ice cores from Antarctica indicates

(A) temperatures and carbon dioxide concentration in the atmosphere have increased. (This has been due to the increased use of fossil fuels resulting in increased emission of greenhouse gases.)

7-18
Scientific modelling allows scientists to
(A) investigate how changes may affect species without actually making those changes. (Modelling allows scientists to predict the effect of possible changes)
7-19
The clearing of land by early European settlers in Australia resulted in
(C) a decrease in biodiversity and extinction of many species. (Clearing land resulted in loss of habit or habit fragmentation. This lead to small isolated gene pools and inbreeding.)
7-20
Which of the following practices could not be used to restore a damaged ecosystem?
(B) introduction of new species (This could alter competition and predator/prey relationships.)

Section II Questions – Suggested Answers

7-21

(a) Plant cell as it contains a cell wall and chloroplasts. Animals do not have these structures.

(b)

	structure	function
A	**ribosome**	site of protein synthesis
B	mitochondrion	**major site of ATP production**
C	**chloroplast**	Site of photosynthesis

(c) C, the chloroplasts, would be the most obvious as they are large, green and probably moving.

(d) Any one of:
- stains disguise the natural colour of organelles
- stains kill the cell.

7-22

(a) Phospholipids: provide the major structure of the membrane and are the site where lipid soluble and small uncharged molecules pass across the membrane. Protein: site where ions diffuse, facilitated diffusion and active transport occur.

(b) The concentration of sucrose is the independent variable in this experiment.

(c) It is important that the Petri dishes were placed next to each other in a cupboard so that the only difference between the dishes was the sucrose concentration. If some other factor for example presence of light was different, the results may be due to the absence of light and not the sucrose concentration.

(d) Cell A would be expected to be from 0.0 M solution, cell B from 1.5 M sucrose solution and cell C from 0.3 M sucrose solution.

(e) Water has moved out of cell B by osmosis. There is a greater concentration of solute outside the cell than in, so water has moved across the semi permeable membranes of the vacuole and cell membrane resulting in the movement of the cell membrane away from the cell wall.

7-23

(a) Enzymes are proteins that speed up specific chemical reactions. They are produced by living organisms.

(b) Enzyme A. The enzyme must be able to breakdown pesticides in the natural environment therefore must have an optimum temperature close to the average environmental temperature. The average temperature of the environment is more likely to be closer to 18°C than 39°C.

(c) No. According to the graph at temperatures above 39°C, the enzyme denatures. This means the shape of its active site is permanently altered so even if the temperature is decreased the enzyme will not function.

(d) No. Enzymes are specific to specific reactions. An enzyme speeds up the reaction where pesticides are broken down. Pesticides have a different chemical structure from that of oil.

7-24

(a) $C_6H_{12}O_6 + 6O_2 + 36\ P_i + 36\ ADP \longrightarrow 6CO_2 + 6H_2O + 36\ ATP$

(b) The energy released in respiration allows the cell to grow, divide, repair, move, synthesise new molecules, active transport, signal and maintain cell structure.

(c) Inputs to photosynthesis: water, carbon dioxide and light.
Outputs of photosynthesis: oxygen, glucose and water.

(d) More chloroplasts would be expected to be found in palisade cells than spongy mesophyll cells. Any one of the following reasons:
- palisade cells are the major site of photosynthesis
- palisade cells are exposed to more light than spongy mesophyll cells
- the rectangular shape and larger size of the palisade cells than the spongy mesophyll cells.

(e) When guard cells take up water they swell and open a pore in plant leaves. This allows carbon dioxide required for photosynthesis to diffuse into the leaf and oxygen produced in photosynthesis to diffuse out of the leaf. Water also diffuses out of the pore, drawing water up the plant in the xylem. When the plant loses too much water the guard cells lose water and the pore is closed, reducing further water loss.

(f) Oxygen is produced in photosynthesis therefore the rate that it is produced can be a measure of the rate of photosynthesis.

(g) Net photosynthetic rate. Plants photosynthesise in the light and respire all the time. Oxygen is produced in photosynthesis but is used in aerobic respiration. The oxygen given off by the plant is the excess oxygen that has not been used in aerobic respiration.

(h) The graph levels off as another factor is limiting the rate of photosynthesis. This factor could be number of chloroplasts or amount of carbon dioxide.

(i) Nitrogen is added to carbohydrate to make protein.

7-25

(a) Plants can make their own organic molecules in photosynthesis whereas animals cannot. Animals obtain organic molecules from the food they eat. Usually the molecules in food are too big to pass across the membranes of the cells lining the digestive tract. They must be broken down or digested into molecules small enough to pass through the membrane.

(b) Mechanical digestion is the physical breakdown of food. It involves muscles or teeth and the product is small pieces of the original food. The food is not chemically altered. Enzymes are involved in chemical digestion and the food is chemically broken down into smaller molecules.

(c) Most mechanical digestion would occur in the first chamber of the stomach. The stomach is muscular allowing for mechanical digestion as the muscles contract and the presence of stones would increase mechanical digestion.

(d) Proteins would be chemically digested in the second chamber. The digestive juices in the second chamber are very acidic. Many proteases function best at a low pH. Proteases digest protein.

(e) In closed circulatory systems, exchange between the tissues and the blood occurs at the capillaries.

(f) In a four-chambered heart, oxygenated blood is directly pumped to the tissues. In simpler hearts, blood is pumped to the lung capillaries and then to the tissue under less pressure. In four-chambered hearts, oxygenated blood and deoxygenated blood do not mix whereas in simpler hearts (three chambered) they do, reducing the oxygen available to the tissues.

7-26

(a) Any two of the following:
- low temperature
- high winds
- high light intensity
- icy substrate.

(b) 1 physiological
2 structural

(c) In the ancestral penguin population, there was genetic variation in wing structure that resulted in a range of wing shapes. Those individuals with wings that resembled flippers were able to swim better, catch more fish, become stronger, attract mates and reproduce. Their offspring inherited the flipper-shaped wings. Those individuals that did not have flipper-shaped wings were poorer swimmers, so they caught fewer fish, became weaker and reproduced less. Over many generations, penguins with flipper-shaped wings became more common.

(d) Convergent evolution. Convergent evolution is when organisms that do not have a recent common ancestor, have similar characteristics because they have been subject to similar selection pressures. Divergent evolution is when organisms with a recent common ancestor differ because they have been subject to different selection pressures. Penguins are birds, dolphins are mammals and sharks are fish therefore they do not have a recent common ancestor. They all swim in seawater therefore are subject to the same selection pressure.

7-27

(a) (i) chimpanzee
(ii) bread mould

(b) Cytochrome c is found in all eukaryotic organisms. After organisms diverge from their common ancestor, mutations in the DNA coding for cytochrome c occur. The more distant the common ancestor, the greater the time for mutations to accumulate.

(c) No. The differences may not be at the same point, i.e. the 12 differences between horse and human cytochrome c may be at different places to the 12 differences between pigeon and human cytochrome c. To determine the relationship between horses and pigeons you would need to compare cytochrome c from the horse with cytochrome c from the pigeon.

7-28

Organisms	Relationship
Two goats in a paddock	Intra-specific competition
Fungi and algal components of a lichen	Mutualism
Wedge-tailed eagle and rabbit	Predator/prey

7-29

(a) The producer in this food web is the plant with roots growing in soil.

(b) The source of energy for the organism(s) named in (a) is the sun.

(c) Organisms at the second trophic level are: soil bacteria, fungi and root feeding roundworms.

(d) The largest trophic level by biomass would be the first – the plants and smallest – either the birds or small animals.

(e) Producers support the rest of the community and on average only 10% of the energy available at one level is passed to the next. Therefore, producers generally have the highest biomass and higher trophic levels the smaller biomass.

(f) Decomposers – fungi and bacteria are present but are not shown as such in the food web. If included there would be further links, e.g. arthropods to fungi and bacteria.

7-30

(a) Estimated population of this plant in the whole area -
Total in quadrats = 4+7+11+9+4+9+7+9+10+7 = 77 in 10 quadrats
This is a 1/10 of the area so multiply by 10 for estimate of total population.
77 × 10 = 770 – estimate of daisy plants in the 100 square metre area.

(b) Average population density – is determined by dividing Total Population by the Total Area = 770/100 = 7.7 daisies per square metre.

(c) Any one of water, soil nutrients, soil water holding ability, light, slope

(d) Solution in table

	6 months before	6 months after
size of 1st sample	195	55
size of 2nd sample – week later	250	65
Number marked in 2nd sample	50	15
Calculated estimate	**975**	**238**

(e) The virus caused a 75% death rate – that is approximately 25% survived. 238/975 = 24.4%

7-31

(a) The southern beech evolved after Africa separated from Gondwana but before the separation of Australia, New Zealand, New Guinea and New Caledonia.

(b) The presence of fossil pollen indicates that the climate of Antarctica was one warmer – temperate.

(c) Any condition that reduces scavenging and decay by other organisms will assist fossilisation.
e.g. **Quick burial.** Fossils are most likely to occur in an environment where there is rapid burial of the dead organism.
– **Cold temperatures**. Low temperatures protect the remains from decomposers.
– **Sap**. Sap also protects organisms from decay.
– **Anoxic environments**. The remains are protected, as aerobic decomposers cannot survive in anoxic environments.

(d) Any one of the following:

Evidence	**Description**
Matching continental margins - shape - type of rock	The shape of the southern continents fit together like a jigsaw. Where the pieces would have met there is **similar geology** (matching tillite-coal layers, dolerite formations in Tasmania and Antarctica, basalt in South America and Africa) suggesting that these areas were formed in the **same place** and at the **same time**.
Position mid-ocean ridges	Mid ocean ridges have been found as predicted where plates are moving apart.
Spreading zones between continental plates	The ocean floor rock increases in age the further it is from the mid-ocean ridges. This indicates that the plates are moving apart, and new rock is forming between them.

Notes

Notes

Notes

Notes

Notes

Notes

Notes

Notes

Notes